Power Cable Installation Practice

Power Cable Installation Practice
E W P Jones, CEng, MIEE

NEWNES

Newnes
An imprint of Butterworth-Heinemann Ltd
Linacre House, Jordan Hill, Oxford OX2 8DP

PART OF REED INTERNATIONAL BOOKS

OXFORD LONDON BOSTON
MUNICH NEW DELHI SINGAPORE SYDNEY
TOKYO TORONTO WELLINGTON

First published 1993

© Butterworth-Heinemann Ltd 1993

All rights reserved. No part of this publication may be reproduced in any material form (including photocopying or storing in any medium by electronic means and whether or not transiently or incidentally to some other use of this publication) without the written permission of the copyright holder except in accordance with the provisions of the Copyright, Designs and Patents Act 1988 or under the terms of a licence issued by the Copyright Licensing Agency Ltd, 90 Tottenham Court Road, London, England W1P 9HE. Applications for the copyright holder's written permission to reproduce any part of this publication should be addressed to the publishers

British Library Cataloguing in Publication Data
Jones, E. W. P.
  Power Cable Installation Practice
  I. Title
  621.319

ISBN 0 7506 1165 0

Typeset by Vision Typesetting, Manchester
Printed and bound in Great Britain by
Biddles Ltd, Guildford and King's Lynn

Contents

Preface vii

1 Insulating materials 1
Introduction 1
Early developments in power cables 2
Types of insulating materials 3
Development of synthetic rubbers and polymers 11
Factors affecting the choice of a dielectric 16
References 18

2 Conductors and cable protection 19
Conductors 19
Cable protection 21
Reference 24

3 Design and planning of power cable systems 25
Direct laying 26
Installation in ducts 27
Installation in racks, cleats or tray 27
Factors affecting power cable supports when erected in air 30
Reference 32

Contents

4 Current rating and cable losses 33
   Ascertaining current rating 34
   Voltage drop 42
   Reference 45

5 Short-circuit performance of power cables 46
   Calculation of short-circuit current 59

6 Power cable accessories 66

7 Power cable jointing 78
   Paper-insulated lead-sheathed cables 78
   PVC- and polymeric-insulated cables 86
   Handling power cables 94

8 Cable testing and fault location 97

9 Earthing and bonding methods 108
   Bonded cable systems 111
   Earth testing 116

10 Assisted-type cables 119
   Oil-filled cable systems 121
   Gas pressure cables 123
   References 125

Index 126

Preface

In his introduction to Hunter and Hazell's *Development of Power Cables*, published by Newnes in 1956, Sir Vincent De Ferranti, son of the famous Sebastian, starts by saying that it was said, facetiously, of the medical profession, that they buried their mistakes. He went on to say that the opposite applied to power cable engineers: their successes remained underground providing the service for which they were designed whilst their failures were dug up and exposed to analysis and publicity.

Having been involved in the transmission of electrical power all of my professional life I have seen such examples of both success and failure and have often thought that there was a need to provide a short practical guide on power cables, together with their installation, as an aid to engineers and technicians. There are a number of excellent books available that cover the design and manufacture of power cables, also embracing the theoretical aspects of this field, so that I formed the view that there was a positive requirement for a relatively concise reference book which would enable the engineer who came into contact with power cable installations only infrequently to have a ready source of information always available on the subject; it is my hope that this book goes some way to meeting this need.

I am greatly indebted to everyone who helped me in the preparation of this book, particularly to Brian Roberts and my former colleagues at BICC Cables, Power Division, to Cellpack

Preface

(UK) Ltd at Hungerford, Langley Engineering of Reading, the 3M Electrical Product Group, Henry A Patterson and Partners of High Wycombe, Dr Jack Taylor of Rotunda plc, Sheen Equipment of Nottingham and the Institution of Electrical Engineers for permission to refer to sections of their Regulations for Electrical Installations.

Finally, my grateful thanks to James Gaussen, formerly of Butterworth-Heinemann, without whose encouragement this book would not have been written.

E.W.P.J.
January 1992

ONE
Insulating materials

Introduction

Nowadays the term 'power cable' is applied to any multicore connection, semi-rigid or flexible, linking the supply point to the apparatus or equipment which it serves.

It is the purpose of this book to deal specifically with the installation requirements of the class of cables which, in the days of the Cable Makers' Association (CMA), were defined as 'mains cables'. Essentially the CMA, or more particularly that part of the CMA organization which dealt with paper-insulated cables, that is the Mains Cable Makers' Association (MCMA), established a bench-mark between cables usually employed for general wiring purposes and those primarily used for electricity supply and industrial applications. The dividing line was set at a conductor size of 0.0225 square inch (14.52 m^2), being decided on the basis that the electricity supply authorities adopted a two-core paper-insulated lead-sheathed and armoured cable of this cross-sectional area as the minimum size of cable to be used in connecting domestic premises to their distribution system – the so-called 'house service cable'.

Until fairly recent years, impregnated paper has been the preferred insulation medium for power cables, having superseded vulcanized bitumen and rubber which were its main competitors in the early days of this century. However, the development of synthetic materials such as the polyvinyl chloride compounds, polyethylene and the various elastomeric compounds has now reached the point where the traditional position of the paper-insulated cable has been severely eroded even at voltages of the super tension category.

The wider choice of insulation media today complicates the design philosophy when preparing a scheme which will incorporate or need to be compatible with an existing system where other insulating and sheathing materials are in use. It is important that the new design parameters are maintained within the original overall design matrix. For this reason it is necessary that this book deals with the design and installation techniques relevant to power cable systems having structural features which, at this point in the twentieth century, may be considered to be outmoded and obsolete.

Early developments in power cables

In the early nineteenth century experimenters such as Michael Faraday had to resort to materials like string and rag to insulate the conductors associated with their experimental apparatus and it was not until the electric telegraph came to the fore that serious consideration was given to insulating conductors which were to be used over relatively long distances.

One of the first materials to be adopted for this purpose was gutta-percha. This natural product, a coagulated latex, found in Malaysia and the East Indies, could be readily processed when heated but assumed a tough horn-like texture at normal temperatures. Being waterproof it soon found a ready use in submarine telegraph cable applications.

By 1849 gutta-percha was in general use and suitable machinery had been developed to mass produce cables insulated with this material. However, on land, gutta-percha suffered from two defects: it was susceptible to oxidation and softened and deformed at moderately high temperatures. Because of these limitations, cable makers turned their attention to natural rubber which, when mixed with sulphur and other ingredients, could be 'cured' or vulcanized by being heated to a temperature of 150°C for a specified period; after this treatment the resultant product, vulcanized rubber, possessed good electrical characteristics together with resiliency and elastic properties.

A further development in this period, originating in the USA, was a system of jute-lapped conductors impregnated with oil, drawn into iron pipes, which, in turn, were filled with oil.

By 1880, with the development of the incandescent lamp, which brought the need to transmit higher levels of power to meet the increasing demand, one arrives at the point at which development of the power cable, as we know it today, may be said to begin.

Types of insulating materials

Vulcanized bitumen (VB)

The demand for a cheap and easily applied power cable insulant led W. O. Callender, a manufacturer of road-surfacing materials, to experiment with the bitumen he imported from the Trinidad lake as a cable-insulating and -sheathing material.

In its crude state bitumen is a dark brown earthy-looking substance, fairly hard at temperatures of 16°C to 20°C and containing about 35 per cent of mineral matter. Callender evolved a process whereby the crude bitumen, together with cotton-seed pitch and an admixture of sulphur, could be vulcanized, thus producing a material which exhibited qualities somewhat similar to those of rubber and gutta-percha.

The finished cable was waterproof, exceedingly stable, and particularly inert to chemical action of an acid nature. The material was, however, susceptible to the action of alkalis although even with these the process of penetration of the sheath was quite slow.

Vulcanized bitumen found much favour for colliery work where it was often exposed to the effects of underground water, frequently contaminated with various chemical salts of both an acid and alkaline nature. Mechanical protection was provided by the application of armouring wire or tapes.

It was not usual to employ VB cables at working pressures above 3300 V and by reason of its propensity to softening of the insulation when in a warm environment, or in a situation of overloading, the cable needed to be derated accordingly:

Air temperature (°C)	Cable current reduction factor
32	0.82
38	0.58

It is interesting to note that at the time of writing (1989), the writer is aware of a 3.3 kV VB installation still operating satisfactorily underground in a Warwickshire colliery.

Varnished cambric (VC)

This material was generally known as 'Empire Cloth' and consisted of a textile fabric impregnated with an insulating varnish. Usually the fabric was cotton but, in latter years, 'Terylene' (polyethylene terephthalate – VT) impregnated with a synthetic varnish found favour in industrial applications where high ambient and elevated operating temperatures were encountered.

The conventional VC cable insulation was manufactured from Egyptian cotton and impregnated with a varnish which consisted of pure boiled linseed oil mixed with gum resins and 'driers' which were allowed to oxidize after being applied to the fabric.

The material was required to have a smooth surface free from any blemish which could impair the varnish film and was subjected to sizing and calendering operations to achieve a satisfactory finish.

The fabric was impregnated by being passed through a varnish bath and then through a series of drying chambers before being hung up in vertical lengths to complete the oxidation process, after which it was cut up into strips and supplied to the cable maker in tape form.

Varnished cambric cables rely entirely upon the varnish to provide the dielectric strength, the fabric being solely the medium to which the varnish film can be attached to provide the insulation covering for the conductor.

In practice two varieties of VC material may be encountered, either yellow, this being the natural form, or black, this colour being produced by the introduction of a bituminous substance.

During manufacture the impregnated tape is applied to the conductor by conventional lapping techniques. The use of an oily compound during this process improves the flexibility of the cable and also helps to exclude any air pockets created during the tape-lapping process.

Because it is not possible to exclude, completely, air and moisture during the manufacturing process, it is important that the cable should be dried as much as possible before use. However, varnished cambric material is *moisture resisting* if not completely moisture-proof and was often enclosed in a lead sheath similar to that used in impregnated paper cables.

The dielectric strength of a VC cable is very much lower than its impregnated paper equivalent but the maximum permissible operating temperature, which is allied to the current-carrying capacity, is quite high, being in the order of 80°C for cables manufactured for service up to 6600 V operating conditions.

The advantages to be gained by the use of VC as against impregnated-paper-insulated cable may be listed as follows:

1 The absence of resin oil avoids problems associated with compound migration and bleeding occurring at terminations

which can be encountered under certain installation conditions.
2 Compound-filled terminal boxes are unnecessary.
3 Compatibility with paper-insulated cable performance simplifies design and installation parameters.

For indoor applications, terminations were usually taped and varnished overall.

It was usual when multicore cables were required to have them laid up with pre-impregnated jute fillings with an overall extruded lead sheath. The use of Terylene tapes with an applied silicone varnish enables VT cables to be operated at temperatures in the order of 120°C.

Impregnated paper

It was Sebastian de Ferranti who, in 1889, took the revolutionary step of insulating copper conductors with brown rag paper soaked in ozokerite, a form of paraffin wax, for his 10 000 V tubular main between Deptford and London's West End.

Subsequently, Ferranti joined the newly formed British Insulated Wire Company at Prescot in Lancashire who had secured the British manufacturing rights for the helical paper-lapping process which had been developed in the USA by the Norwich Wire Company; thus the use of impregnated paper had become well established by the start of the twentieth century.

Impregnated paper remained unchallenged as the principal insulating medium for power cables until relatively recent years by reason of its good dielectric properties as well as economic considerations. Manufacturing techniques were developed and refined over the first half of this century as indeed were the dielectric and mechanical properties of the paper insulating medium itself. Indeed, by the 1950s typical cable-insulating paper had a DC or impulse strength of 1500 kV/cm with long-term AC strength exceeding 300 kV/cm and a short-term overvoltage withstand of 500 kV/cm.

The paper used for insulating power cables must have mechanical properties which will allow it to be applied helically to a conductor at relatively high speeds and possess physical and chemical stability to meet the wide range of conditions it will encounter both during manufacture and in service.

Generally, wood-pulp, manilla and cotton, or mixtures of these, are satisfactory in service. In the main, wood-pulp has, in recent years, been the principal material used by cable makers.

Both single and two-ply papers are used but, generally, two-ply paper is preferred, as the possibility of surface defects in the paper causing breakdown is much reduced.

The electric field in a high-voltage cable having circular conductors is radial, the dielectric closest to the conductor being subjected to the highest field strength, the field strength diminishing outwardly; therefore the cable maker will vary the thickness of individual papers in order to grade the dielectric to suit the levels of electrical strength encountered through the layers of paper from the conductor outwards. The relation between the thickness of the impregnated paper tape and its electrical strength is shown in Figure 1. It is also possible to show (Table 1) the magnitudes of electrical strength on certain insulating papers of different thickness.

It will be seen that by using the tabulated figures relating to paper thickness against the electrical strength in kilovolts/centimetre it would be possible, in theory, when designing a cable suitable for operation at 132 kV, to reduce the dielectric thickness from 10.414 mm, using 4 mil paper only, to 6.350 mm using a combination of 1.0, 1.5 and 2.5 mil papers.

However, it has to be pointed out that when applying such very thin papers in the standard manufacturing process there would be considerable engineering difficulties.

Readers should note that the term 'mil' tends to be used by cable makers and relates to a measurement of 1000th of an inch or 0.0254 millimetres.

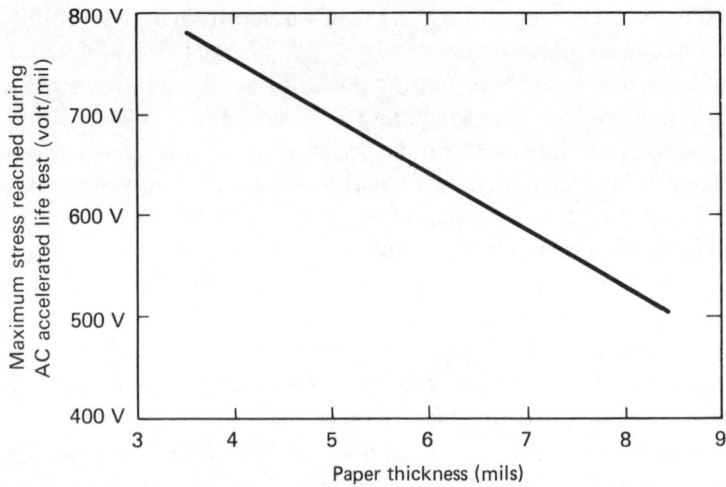

Figure 1 *Breakdown strength related to paper thickness (1 mil = 0.0254 mm)*

Table 1 Effect of paper thickness on electric strength

Thickness (mils)	Electric strength (kV/cm)	
	Short time AC RMS	Impulse Peak
1	850	2000
$1\frac{1}{2}$	700	1800
$2\frac{1}{2}$	580	1700
4	500	1550
7	400	1250

Impregnants for paper cables

Early paper cable used rosin oil as an impregnant but, by about 1912, mineral oils were introduced and these brought about a great improvement in dielectric performance.

Insulating materials

Nowadays refined mineral oil to which purified gum rosin has been added to increase the viscosity is the general impregnant for standard usage. Although the addition of the rosin reduces the mobility of the impregnating compound the problem of compound migration under vertical or steeply sloping installation conditions remains and this difficulty led to the development of mass impregnated-non-draining compounds (MIND cables).

Initially, cable makers approached the problem from several

Figure 2 *185 mm^2 three-core cable, paper insulated, lead sheathed, single wire armoured and PVC sheathed overall: 11 kV MIND to BS 6480 (BICC Cables Ltd, Power Division)*

Power Cable Installation Practice

directions but, nowadays, the use of microcrystalline waxes, polyethylene, polyisobutylene together with rosin and added to the mineral oil has the effect of turning the impregnant into a jelly-like material with considerably enhanced viscosity characteristics.

Prior to the introduction of MIND cables great difficulties were encountered with conventional oil/rosin-impregnated cables installed between different levels, the impregnant having the propensity to migrate to the lower levels, causing voids at the higher end, leading to subsequent electrical failure whilst, at the lower end, the lead sheath of the cable was subjected to excessive pressures, causing it to distend, leading to fractures and leakage of the impregnant from the lower termination, often causing damage to connected equipment such as switchgear.

These problems can be avoided by the use of MIND cables as the impregnant will remain *in situ*; however, care should be taken when handling these cables at low temperatures as they become slightly stiff and need to be handled carefully when bending under cold installation conditions.

For the record, MIND cables have been used successfully all over the world for the past 40 years and have a very satisfactory service record up to and including 33 kV.

Rubber as a power cable insulant

Early power cable installations made extensive use of natural rubber but, with the introduction of high-voltage AC circuits, there was a need to seek a cheaper material and also a dielectric which possessed a lower permittivity.

However, the flexibility and toughness of rubber ensured that in some fields, such as mining, it was invariably preferred to other dielectrics.

The conversion of natural rubbers into a suitable form for use as a cable insulant needs the addition of sulphur and other ingredients which require to be mixed together; the resultant mixture

Insulating materials

becomes a compound which is subjected to the vulcanization process which needs the material to be heated for a specified period at temperatures in the order of 150°C.

The resultant substance, vulcanized rubber (VR) is largely non-hygroscopic and although its insulation resistance is not as high as that of pure rubbers, it possesses high levels of resilience and elasticity.

Nowadays it is unusual to encounter vulcanized rubber power cables other than in the mining and quarry fields of application where the excellent mechanical and resilient properties of the material come to the fore. It should be noted that in high-voltage applications vulcanized rubber is susceptible to attack by the ozone which is produced by the action of electrical corona discharge on oxygen or air and which is a powerful oxidizing agent. Attack by ozone causes cracks to appear in the dielectric which quickly brings about failure of the insulation and break-down of the cable.

The problems of cable failure resulting from the effects of ozone led cable makers to the development of 'ozone-proof' rubber cables where the vulcanized rubber was protected by ozone-proof tapes which incorporated petroleum or bituminous compounds. In addition, care was taken at termination to incorporate stress relief components to reduce the risk of discharge.

Development of synthetic rubbers and polymers

Since World War II considerable development has taken place in the field of man-made products and these materials have now become the major manufacturing elements in power cable production.

Details of the compounds generally used in power cable manufacture are briefly given here; readers who require more specific information are recommended to consult the more detailed publications which are available in this field.

Butyl rubber

Butyl rubber is a copolymer of isobutylene and isoprene.

The process of copolymerizing produces a material which may be compounded and vulcanized in a similar way to natural rubber but has a higher resistance to oxidation and is capable of being operated indefinitely at temperatures in the order of 80–85°C. In the early 1960s butyl rubber compounds having low water absorption levels were developed to investigate the possibility of dispensing with metallic sheaths for direct burial in the earth of power distribution cables (Barnes, 1964).

Polyvinyl chloride (PVC)

In its polymer form this material is a white powder which is compounded by mixing with plasticizers, fillers, stabilizers and lubricants. After heating for periods of 2–10 min at temperatures in the order of 150–170°C the mass gels and becomes rubber-like at this high temperature. It becomes possible to alter the properties of the compound by varying the nature and quantity of the plasticizer.

Generally, PVC compounds have permittivities of between 4.0 and 6.0 and power factors between 0.05 and 0.1, providing adequate electrical properties up to 6.0 kV.

PVC does not have the flexibility of rubber and will tend to flow under conditions of elevated temperature. It tends to become stiff and brittle at low temperatures but it is possible to compensate for this by adopting a suitable compound formulation which will provide the properties required for a specific application.

It has the advantage over rubber of being inert to oxygen and resistant to oil, weather and chemicals generally.

Compounds of PVC have been developed which have modified burning characteristics, thus resisting flame propagation. It is also possible to reduce the level of hydrogen chloride gas emission by the addition of a specially processed calcium carbonate filler.

Insulating materials

Figure 3 *95 mm² four-core cable, PVC insulated, single wire armoured and PVC sheathed overall: 600–1000 V class to BS 6346 (BICC Cables Ltd, Power Division)*

Dedicated low-smoke and fume cables are available for specific applications.

Polyethylene

Polyethylene is obtained by the polymerization of ethylene at pressures of about 1500 atmospheres.

The advantages of polyethylene over PVC are improved stability (plasticizer not required) and low electrical losses, but the material has a propensity towards oxidation produced by ultraviolet radiation and requires the addition of carbon black to the compound if the cable is to be exposed to sunlight.

Polyethylene has only been used in the UK to a limited extent as a power cable insulant. Although the electrical characteristics of polyethylene are excellent, its use as a cable insulant has been largely limited by its relatively sharp softening point near 100°C. For this reason the upper operating temperature is limited to 70°C.

Crosslinked polyethylene (XLPE)

The superior electrical properties of polyethylene only found a use in relatively restricted fields by reason of the low melting point of the polymer, but the technique of 'crosslinking' has enabled this deficiency to be overcome.

The process of crosslinking is generally achieved by introducing peroxides into the polymer and subjecting the resultant material to steam heat under pressure.

Recently, a new technique for crosslinking has been introduced and is known as the 'monosil' process (McAllister, 1982). The crosslinking process involves the use of silicones and has a simplified extruding operation. The monosil process is in general use for cables manufactured in the 10–30 kV range.

XLPE will operate continuously at temperatures in the order of 90°C and has a short-circuit rating which will accept 250°C as a final conductor temperature.

Ethylene propylene rubber (EPR)

This material has, in recent years, tended to supersede butyl rubber by reason of its improved performance.

The product was first developed in Italy during 1961 by the

Insulating materials

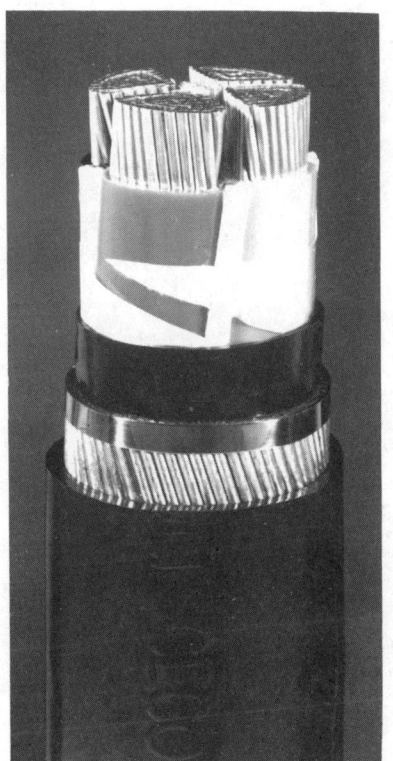

Figure 4 *500 mm² four-core cable. XLPE insulated, single wire armoured and PVC sheathed overall: 600–1000 V class to BS 5467 (BICC Cables Ltd, Power Division)*

Montecatini company of Milan who produced an ethylene-propylene copolymer which was assigned the trade name 'Dutral' (Black, 1983). This material was found to have rubber-like properties and could be cured with either a sulphur or peroxide catalyst.

It was soon to find favour with cable makers, having, like the other thermoset material, crosslinked polyethylene, excellent

15

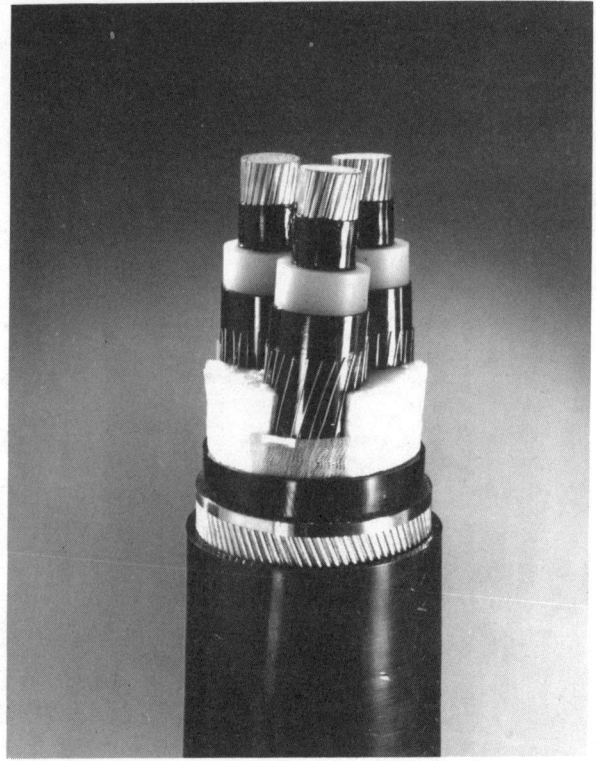

Figure 5 *300 mm² three-core cable, XLPE insulated, single wire armoured and PVC sheathed overall: 11 kV to Spec. IEC 502 (BICC Cables Ltd, Power Division)*

electrical and elevated temperature performance together with good flexibility and abrasion resistance.

Factors affecting the choice of a dielectric

Modern practice has all but seen the demise of paper-insulated cable below 11 kV.

Insulating materials

The introduction of the BS 3346 in 1961, now BS 6346, which provides designs for PVC-insulated, armoured and PVC-sheathed cables up to and including 3.3 kV, together with British Coal Specifications 295 and 656, has provided a range of cables up to and including 11 kV. To this must be added IEC 502 which covers XLPE cables up to and including 22 kV. These specifications provide a range of cables which will, under normal conditions of environment and temperature, be adequate for most applications.

It should be noted that British Coal in their Specification 656 have favoured EPR for power cables in the 6.6–11 kV range. Both EPR and XLPE have similar properties but there is a strong body of opinion which favours EPR insulation by reason of its greater flexibility and resistance to crushing, a hazard often encountered in mining and quarry work.

On the other hand, XLPE requires fewer additives than EPR during production and this reduces the risk of contaminants being introduced into the dielectric during the manufacturing process. Additionally, the reducing cost of XLPE makes it likely that, on economic grounds, XLPE will tend to be the dominant insulant in future years. Indeed, BS 5467 provides a range of XLPE power cables for the medium-voltage range, i.e. 600–1000 V and 1900–3300 V.

Whilst it is not within the scope of this book to discuss at length, or in any degree of detail, the relative merits of the various polymeric compounds which are available, it is useful, nevertheless, to identify some of the more salient features of these comparatively new materials.

The UK electricity supply industry was attracted, in the later 1960s, towards the economies which could be achieved in jointing operations by the adoption of polymeric designs, having noted the decline in paper insulation in favour of the new materials, both on the European continent and in the USA.

However, as early as 1967 reports were being received from both North America and Japan of a phenomenon appearing in these designs known as 'water treeing', a manifestation which may be described as a progressive deterioration of the dielectric resulting from defects in the insulation such as voids or inclusive contaminants.

Both EPR and XLPE can develop water trees but the latter would appear to be more susceptible than the former, although it is reported that many XLPE cables have operated satisfactorily despite the presence of extensive water trees.

The provision of moisture-resisting barriers in the cable construction will reduce the risk of trees forming in the dielectric but, obviously, the addition of such safeguards will have an effect on the economics of any particular design.

Although the foregoing relates to high-voltage cables, principally 11 kV, cable manufacturers will seek to standardize on dielectrics so that any extensive move towards a specific polymeric insulation in the high-voltage range will inevitably mean that this particular material will also become competitively available in the low-voltage range. BS 5467 illustrates this tendency in respect of low-voltage XLPE cables.

Much has been written lately regarding the pros and cons of polymeric cables and the reader who wishes to pursue the subject will find a wealth of information and comment in the technical papers published by the Institution of Electrical Engineers and other learned bodies in recent years. However, it should be noted that improvements in the manufacturing process together with additives to compounds is an ongoing factor and manufacturers now confidently predict that today's XLPE cables are increasingly resistant to water tree development and will provide improved reliability and service performance.

References

Barnes, C. C. (1964) *Electric Cables*, Pitman.
Black, R. M. (1983) *The History of Electric Wires and Cables*, Peter Peregrinus.
McAllister, D. (ed.) (1982) *Electric Cables Handbook*, Granada.

TWO

Conductors and cable protection

Conductors

The power cable normally contains a conductor or conductors made up of a number of bare copper or aluminium wires. The conductors may be stranded or solid and can be circular in profile, shaped or tightly compacted depending upon the type of cable and its service requirements. BS 6360 'Conductors in insulated cables' covers the properties and dimensions of the standard configurations used by cable makers.

It should be noted that prior to 1970 the Imperial system of measure was used in Great Britain and conductor size was expressed in units of square inches. The metric system (mm^2) used in continental Europe now applies in the UK and has become virtually international with the exception of North and South America and those countries which follow American standards.

The American system of conductor sizing is based upon the American Wire Gauge (AWG) up to the equivalent of $107\,mm^2$, after which size the 'circular mil' unit of area is used to express the cross-sectional area of conductors. A circular mil is the measurement given to the cross-sectional area of a solid wire 1/1000 inch (0.025 mm) in diameter.

IEC 228 now effectively establishes the standards commonly in use throughout most of the world.

The relationship between the former Imperial range of power cable conductors and the standard metric range is given in Table 2.

Table 2 Conductor sizes of power cables

Standard metric size (mm^2)	Nearest imperial size (in^2)
16	0.0225
25	0.04
35	0.06
50	0.06
70	0.10
95	0.15
120	0.20
150	0.25
185	0.30
240	0.40
300	0.50
400	0.60
500	0.75
630	1.00

When a high-voltage AC conductor carries current it tends to be confined to a thin surface layer; to overcome this problem large conductors are often divided into sector shapes, usually four in number, a layer of insulating paper being lapped over alternate sectors; this type of construction is known as a 'Milliken' design.

Copper conductors in power cables are generally of the stranded variety to provide the necessary flexibility but IEC 228 does cater for solid conductors up to 150 mm^2 should they be required.

The high price of copper during the 1960s led to a move towards aluminium as a conductor material although initially there were problems in jointing this metal due to oxidation. However, techniques involving special fluxes and abrasion soldering methods largely overcame these difficulties, UK Area Electricity Boards being in the forefront of these developments.

The increasing use of cold compression jointing has assisted the

Conductors and cable protection

expanding use of aluminium as a power cable conducting during periods of high copper price.

There are sound economic grounds for the use of solid sectorially shaped aluminium conductors and such a range of cables is provided in BS 6346 and also in BS 5467, being in the 600–1000 V category.

Cable protection

Traditionally the paper-insulated lead-sheathed cable was protected by the application of galvanized steel wires or a double steel tape liberally coated with bitumen.

It was the practice of a number of supply authorities, prior to nationalization in 1948, to specify steel wire armour (SWA) for their high-voltage underground mains and double steel tape armour (DSTA) for their low-voltage distributors. In city and urban areas where cable routes under pavements and footpaths tended to be somewhat congested, together with the then universal practice of serving all cables that were to be laid in the ground with jute, the use of armour to identify the class of cable was an effective safety measure in enabling cable jointers, who worked on low-voltage distributors in a live condition, from inadvertently opening up a high-voltage feeder. Nowadays, of course, the use of colour-coded PVC oversheaths together with the embossing of details of the class and working voltage data on the sheath obviates the need to adopt such a practice.

The lead sheath in a paper-insulated cable exists to protect the insulation from moisture but it, in its turn, needs to be protected from damage, in that penetration of the lead, allowing ingress of water, would quickly lead to cable failure. Bitumen together with hessian tapes and/or jute has traditionally been used as a moisture-excluding barrier and also forms a suitable bedding for the laying up of the armouring.

During the 1950s PVC was introduced as a power cable

sheathing material and now is the most popular type of protective finish in use.

Reference has been made to the use of aluminium as a conductor material; there had been little use of this metal as a sheathing for power cables, by reason of its stiffness and its susceptibility to corrosion, until the late 1960s and early 1970s, when the old technique of swaging down of aluminium tube to form a sheath was superseded by direct extrusion presses which enabled a new form of low-voltage cable known as 'CONSAC', in which the aluminium sheath provided a combined neutral and earth conductor, to be introduced for Area Board distribution systems.

One of the problems encountered when using aluminium-sheathed cables was the lack of flexibility compared with the conventional paper/lead-armoured cables; however, corrugated aluminium sheath protected by an extruded PVC oversheath is now well established as the standard protection for paper-insulated 11 kV cables.

PVC oversheaths, when applied over conventional armour, are usually extruded over a layer of bitumen-impregnated bedding tape of a fibrous nature. There is always a likelihood of the oversheath being penetrated and allowing the ingress of moisture; to combat this possibility a liberal application of bitumen forms a barrier between the inside face of the oversheath and the cable sheath proper.

Bitumen is not normally applied directly to the wire armour under extruded sheaths, but in mining-type power cables, where the risk of mechanical damage and consequent water penetration is particularly high, it is applied freely to the double wire armour (DWA) cable which is generally installed underground.

As has already been mentioned, extruded sheaths can be provided which restrict flame propagation and reduce the evolution of toxic gases; the reader is referred to BS 6425 Part 1 and IEC 754 Part 1, for further information on this subject.

Damage to protective sheathings frequently arises from attack by rodents or insects such as termites, ants and teredo worms.

Conductors and cable protection

Attack by these insects is mainly confined to tropical and subtropical regions. Termites feed on cellulose, usually obtained from wood, but will attack cable sheaths if encountered. Ants are usually found in sandy soils but can generally be controlled by treatment of the ground in proximity to the cable by insecticides such as DDT. Treatment of the servings and sheathing material with contact insecticides, whilst eventually reducing the number of insects in the long term, does not prevent penetration of the sheathing by the creatures in the short term. The most effective method of protection is by the incorporation of a protective metal tape, generally bronze or brass, which will withstand the predations of these pests.

Reference has been made to protective finishes in the form of servings and extruded sheaths; prior to the widespread adoption of extruded sheaths, power cables installed indoors were frequently protected by a textile braid which was generally treated to provide a fire-resisting finish.

Power cables in the UK, when laid directly in the ground, require to be identified and indeed protected against damage from excavation carried out on adjacent utilities or buildings. It is usual for cables to be laid in a trench between 500 mm and 800 mm in depth (depending upon voltage) and initially covered with sand or riddled soil to a depth of about 50 mm above the cable, after which earthenware or concrete cover tiles are laid on the bedding material which has been compacted to form a flat surface.

Nowadays it has become the practice in the electricity supply industry to dispense with the warning tile on economic grounds and in its place to lay a plastic tape carrying a warning message. It is the writer's view that this is a false economy, as the concrete or earthenware cover tile provides the first line of defence in protecting the cable from penetration by sharp objects.

It is still possible to find, on old underground installations, protection being provided by warning boards; these consist of 10 ft (3.05 m) lengths of creosoted timber laid in a similar manner to warning tiles.

It may also be necessary to protect power cables by drawing the cables into pipes or ducts and here it is necessary to remember that the annular space between the cable and the inside surface of the duct restricts the transfer of heat and will reduce, significantly, the current rating of the cable. It is possible to overcome this problem by introducing into the duct a mixture of bentonite, cement and sand to which water has been added, thus producing a grout which remains in a semi-liquid state, the ends of the duct being sealed to prevent escape of the mixture (McAllister, 1982).

Apart from the use of suitably compounded sheathing materials to inhibit the spread of fire it is also possible to provide suitable protection by painting the cable *in situ* with suitable fire-resisting paint.

The IEE Regulations for Electrical Installations provides guidance in selecting suitable forms of protection for power cables when they form part of an electrical installation.

Reference

McAllister, D. (1982) *Electric Cables Handbook*, Granada.

THREE
Design and planning of power cable systems

A power cable installation will usually consist of a distribution system centred upon one central or several major substations at which the incoming supply is transformed to provide an appropriate voltage to meet the requirements of the consumer or consumers.

The choice of voltage is dependent upon a number of factors but generally factory and urban distribution systems operate at 11 000 V, with substations located at load centres providing medium-voltage distribution at 415 V three-phase and neutral for power and lighting purposes. Where large motors are required, in industrial applications, it may be advisable to feed these directly from the 11 kV system or provide a 3.3 kV supply at the appropriate works substation.

Most medium-voltage substations will have transformer capacity not exceeding 1000 kVA, to restrict short-circuit currents, and can be fed either from radial feeders emanating from the major substation or by a ring main system which provides an alternative supply in the event of cable or switchgear failure.

There may be sound economic reasons to consider a radial system with duplicate feeders as an alternative to a ring main system but such judgement will depend upon a study of the various factors involved in a specific scheme.

Most industrial installations require cables to be run in a variety of ways, both indoors and outside buildings; the choice of protective finishes must take into account the environmental conditions and risk of mechanical damage which may occur when power cables are installed in exposed overground situations.

Generally speaking, the most satisfactory method of installation is to bury the cable directly in the ground or have it drawn into ducts or pipes, but it must be remembered that in the case of duct installations the current-carrying capacity of the cable suffers by reason of the high thermal resistance of the air surrounding the cable. It is, however, possible to improve the current-carrying capacity of a power cable installed in ducts by methods described elsewhere in this book.

There are advantages in having a power cable system completely contained within a duct network in that excavation of cables or dismantling of cleating systems is avoided when repairs or modifications are necessary. Similarly, spare ways may be included in the initial installation so that extensions or increases in capacity may be carried out with the minimum of disruption, a very important factor in a busy industrial plant. On the other hand the higher initial cost must be weighed against that of a direct-laid or cleated system which, if properly selected for the operating conditions and having an element of additional capacity to allow for load growth, will provide an acceptable level of service over the life of the installation.

Basically, then, the designer has the choice of three installation systems:

1 Laid directly in the ground;
2 Drawn into ducts or pipes;
3 Installed on racks, cleats or tray.

Direct laying

Cable laid directly in the ground should, preferably, rest upon a bedding of soft riddled earth free from any hard or sharp substances which may come into contact with the cable; this is particularly important should there be any ashes or other corrosive matter present in the ground, and suitable bedding material should be imported for this purpose if none is immediately available locally.

The route should be selected to avoid obstructions and appropriate locations established to mount the drum holding the cable to be installed so that the cable may be pulled directly into the prepared trench.

Subsequently, the trench will be backfilled and the surface, after consolidation of the backfill has taken place, reinstated to its former condition, e.g. tarmac, concrete paving or turf.

Installation in ducts

The vitrified clay single and multiway duct is still, after very many years, the most popular form of conduit for power cable installation. The individual conduits are joined by means of a self-aligning spigot and socket joint and are manufactured in accordance with BS 65 and BS 540. In the case of a single duct installation, PVC tubing has advantages in that it is available in longer lengths and is flexible. Concrete and reinforcing mesh may be used to surround a multi-duct installation. After installation the duct run should be rodded or drawn through with a mandrel to ensure correct alignment and freedom from obstruction. The duct mouth should be sealed to prevent ingress of soil and a draw line threaded through the conduit run to facilitate the ultimate installation of the cable.

Draw pits should be constructed of brick or concrete and provided with a cover or concrete slab, dimensions can be such as to permit joints to be accommodated in the pits where necessary.

Installation in racks, cleats or tray

Power cables can be provided with a variety of supports which will meet specific requirements. They should be designed in such a manner as to avoid sharp points or edges which can damage the cable; similarly, any changes of direction of the support system should ensure that the cable will not be bent to a smaller radius than that recommended by the manufacturer.

Generally, for power cables insulated with impregnated paper and lead sheathed, with or without armour, and operating up to and including 11 kV, the bending radius should not exceed 12 × outside diameter.

Cables insulated with PVC or XLPE, again with or without armour, should have a minimum internal radius of not less than 8 × outside diameter of the cable when installed.

In the design stages of the installation it is usually possible to group cables to reduce the number of individual runs which, in turn, will facilitate and simplify the cable support system, thus leading to cost reduction in the fabrication of supports and in the economics of installation.

Rack and ladder installations

There are a number of proprietory cable racking systems on the market and these generally consist of a steel support framework to which steel channels, cantilever arms, brackets and other assemblies may be fixed.

Cleats or saddles may be used to attach the cables to the structure and care should be taken to ensure that cleats used to secure single-core AC cables are manufactured from non-magnetic materials to avoid problems with circulatory currents.

Three-phase AC circuits consisting of three single-core cables should be installed in trefoil formation and consideration given in the design stage to the strength of the support and cleating system under short-circuit conditions when high mechanical bursting forces may be encountered.

Cable ladder systems were developed to accommodate heavy cable loads over wide spans and are much favoured in the oil and petrochemical industries. The system consists of straight lengths,

Design and planning of power cable systems

usually about 3 m in length, with rungs approximately 300 mm apart, and a range of T-pieces, bends etc. which enable the designer to produce a compact supporting structure easily assembled on site.

Cleats

This is probably the most-used method of single support for power cables in low-cable-density installations.

A range of designs utilizing steel, aluminium, nylon, moulded polyethylene and glass fibre are available and may be selected according to the environmental and mechanical conditions applying in a specific installation.

Care should be taken in determining spacing of cleats; too close spacing leads to sheath damage in paper lead cables under load-cycling conditions. Similarly, excessive spacing can induce damaging mechanical stresses, again leading to cable damage.

It is important that the cleat sizes selected are appropriate to the diameter of the cable; undersized cleats will produce excessive compression forces on the cable, whilst oversized components will permit the cable to move under loading conditions.

The wide range of materials used to manufacture cleats enables the designer to select a product which will be compatible with the cable-sheathing material and the environmental conditions in which the installation will operate.

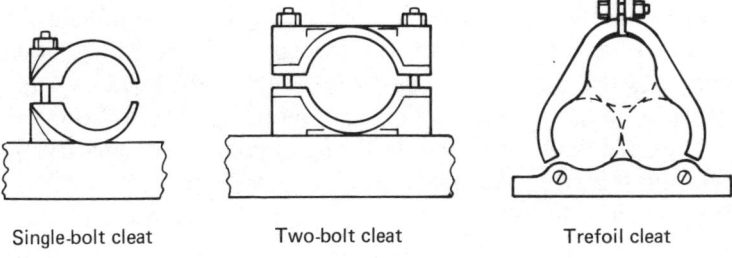

Single-bolt cleat   Two-bolt cleat   Trefoil cleat

Figure 6   *Cleats*

Cable tray

This system has been in use for many years, tray being generally supplied in lengths of the order of 2–3 m with a similar range of accessories to that of cable ladder which has, in recent years, tended to supersede traditional tray for power cable installations.

Tray is formed from plain steel sheets being perforated and finished with zinc chromate paint, or hot dip galvanized for external applications. It can be supplied in various widths.

Support fittings for tray work can be fabricated from mild steel bar and should be treated with zinc chromate or be hot dip galvanized before erection.

Factors affecting power cable supports when erected in air

It is important to ensure that power cables installed in cleats, racks, or the 'J'-type hangers which are sometimes used, are not subjected to excessive movement as a result of expansion and contraction resulting from regular load cycles.

More detailed information than that published by manufacturers relating to the mechanical problems of heavily loaded power cables on spaced supports can be obtained from Holttum (1955).

When groups of multicore power cables are installed in air, adequate spacing should be provided to permit the dissipation of heat; even so it may be necessary to derate some circuits, appropriate details being provided in the IEE Regulations for Electrical Installations and ERA Reports 74-27 and 74-28. Other relevant ERA Reports are 85-0187, 88-0458 and 89-0135. Generally speaking, when spacings in the order of 75 mm to 150 mm can be provided between cables operating at their rated capacity, no derating should be necessary.

Cables installed in tiers should be cleated with spaces to allow air circulation, perforated tray or cable ladder installations, allowing the free flow of air through the cable groups.

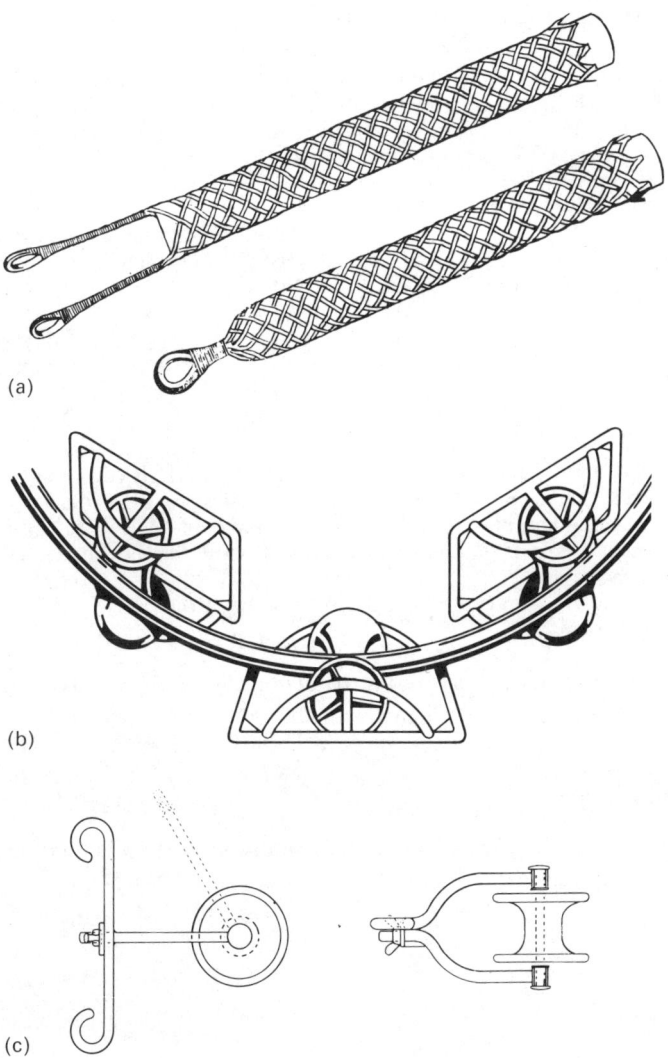

Figure 7 *Power cable installation equipment (Sheen Equipment, Nottingham): (a) cable stocking; (b) cable rollers positioned for corner work; (c) cable suspension roller*

Table 3 Spacing of supports for power cables

Overall diameter of cable (mm)	Non-armoured cable (mm)		Armoured cable (mm)	
	Horizontal	Vertical	Horizontal	Vertical
≤9	250	400	–	–
9–15	300	400	350	450
15–20	350	450	400	550
20–40	400	550	450	600
40–50	600	800	900	1100
50–60	750	1000	950	1100
60–70	900	1200	1000	1200
>70	1000	1400	1200	1400

Notes
1 Based upon Table 11A of the IEE Regulations (15th edn) and, where outside diameter >40 mm, upon data published by BICC plc.
2 Horizontal spacings apply to runs that make an angle >30° from the vertical. For runs that make an angle <30° from the vertical, the vertical spacings apply.

Trefoil cleats are available to accommodate the three single-core phase cables of a heavy-current AC circuit.

Reference

Holttum, W. (1955) The installation of metal sheathed cables on spaced supports, *Proc. IEE*, Part A, **102**, 729–742.

FOUR
Current rating and cable losses

In selecting a suitable conductor size for a power cable the main factors to be considered are:
1 Current to be carried continuously;
2 Permissible voltage drop;
3 Maximum fault current which can occur and its time of duration;
4 Ambient temperature conditions of the installation.

In addition to the above it is necessary to have regard to the prime cost of the installation and to the possibility of the loading being increased by future developments.

The ratings of power cables are readily available in published form in the IEE Regulations for Electrical Installations together with figures published by cable manufacturers in their technical literature.

A comprehensive reference source is IEC 287, entitled 'Calculation of the continuous current rating of cables (100% load factor)'. In addition, IEC 448, 'Current-carrying capacities of conductors for electrical installations in buildings', gives ratings for unarmoured cables under standard conditions.

The former British Electrical and Allied Research Association (now ERA Technology Ltd) has been very active in the field of power cable current-carrying calculation and research, their work being largely covered in ERA Report No. 69-30. Part 1 of this

33

publication provides details of ratings for paper-insulated power cables up to and including 33 kV, whilst Part 3 covers PVC power cables up to 3.3 kV; Part 5 contains details of thermoset cables, again to 3.3 kV. (A thermosetting material is a plastic which does not soften significantly on heating to temperatures below its decomposition temperature, e.g. XLPE.) Other factors affecting ratings are referred to in Chapter 3.

Ascertaining current rating

In operation cables are subject to electrical losses which are manifested in the form of heat, which, being initially generated in the conductors, passes through the various materials which either insulate or protect the cable before being dissipated into the environment surrounding the cable. It is possible to express this process in a manner analogous to that of an electrical circuit and such losses are generally referred to as ohmic losses.

Figure 8 provides a diagrammatic explanation of the mechanics of heat flow in a conventional three-core, paper-insulated, lead-sheathed, armoured and protectively sheathed cable. Using the technique illustrated it is possible to calculate the losses involved and, from that result, derive the permissible current ratings. The mathematical treatment of this subject is fully covered in McAllister (1982) and other works of that nature. Apart from the heat losses described there are other factors which need to be taken into account in determining the safe current-carrying capacity of a power cable; these arise through skin and proximity effects together with losses in the dielectric. Again, detailed mathematical treatment is given in McAllister (1982).

Generally speaking, armouring and sheathing losses in multi-core cables do not have a very important effect on the overall loss figure. However, in single-core cables, losses arising from circulating currents and eddy currents can be significant.

The adoption of sheath bonding when operating single-core metallic-sheathed cables enables sheath losses to be eliminated,

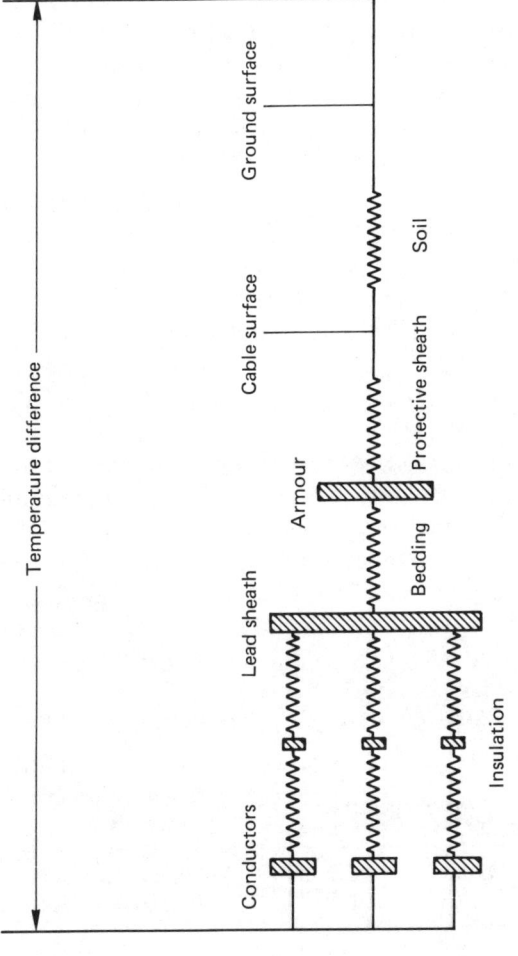

Figure 8 *Heat flow in a conventional three-core paper-insulated cable, represented in circuit diagram form*

the eddy current loss being generally small relative to the sheath losses. The reader is referred to IEC 287 for details of the calculations involved. Installation techniques relevant to bonding single-core cables are dealt with elsewhere in this book.

Note should be taken that single-core steel wire or steel tape armoured cables are not recommended for use in AC circuits due to the heavy losses which occur when magnetic material is used for armouring. When the degree of protection afforded by armouring is required on such circuits, non-magnetic material or aluminium may be used but, nevertheless, allowance for circulating and eddy currents must still be made.

Having established the basic concepts which govern the designed performance of the power cable it is necessary to consider other factors which also have an effect on the rating of the cable.

Temperature

The ambient temperature at which a cable operates varies geographically and, of course, depends on whether the installation is out of doors, within a building or buried in the ground. In the UK the normally accepted temperatures, upon which ratings are based, are 15°C for buried cables, 25°C outdoors in air and 30°C in air within buildings.

Table 4 Ambient air and ground temperatures based upon geographical climate (McAllister, 1982)

Climate	Air temperature		Ground temperature 1 m depth	
	Minimum (°C)	Maximum (°C)	Minimum (°C)	Maximum (°C)
Tropical	25	55	25	40
Subtropical	10	40	15	30
Temperature	0	25	10	20

Current rating and cable losses

Table 5 Conductor temperature limits for standard cable types (McAllister 1982)

Insulation	Cable design	Maximum conductor temperature (°C)
Impregnated Paper 600 V–6.0 kV	Belted	80
Impregnated Paper 6.0–10 kV	Belted	65
Impregnated Paper 6–15 kV	Screened	70
Impregnated Paper 12–30 kV MIND	Screened	65
Polyvinyl chloride	All designs	70
Polyethylene	All designs	70
Butyl rubber	All designs	85
Ethylene propylene rubber	All designs	90
Cross linked polyethylene	All designs	90
Natural rubber	All designs	60

Table 4 gives details of temperature levels which can be expected generally in tropical and semi-tropical zones. Again IEC 287 provides more comprehensive information.

It is, of course, important to take seasonal conditions into account when determining ambient temperature conditions; ratings based upon winter temperature statistics may prove to be inadequate in high summer conditions. Similarly, the choice of insulating, protective and sheathing materials may restrict the heat flow from the conductors to the cable surface as demonstrated in Figure 8.

Maximum cable operating temperatures based upon design and material parameters have been agreed by the IEC and are detailed in Table 5.

Consideration must also be given to the effect of high current loadings, resulting in elevated conductor temperatures, upon the soil surrounding the cable, which can dry out at cable surface temperatures in excess of 50°C, thus allowing the cable temperature to rise to unacceptable levels.

Installation conditions

When cables are buried in the ground the heat transmitted from the current-carrying conductors passes into the surrounding soil. Generally, soils containing a high degree of moisture will have a lower thermal resistivity and therefore provide better heat dissipation properties than porous and well-drained ground.

The soil resistivity (g) varies throughout the year, low values occurring in the spring when the moisture content is high and the temperature low, whilst in the summer the converse occurs. Similarly, the effective soil thermal resistance will tend to be low in areas subject to frequent flooding so that in the design stage of large power cable schemes it is often desirable to determine the thermal resistivity at various points on the cable route.

Ground temperature is also important and the reader is once again referred to IEC 287 for more detailed information in respect of the various countries of the world.

Table 6 gives general guidance for values of g expressed in kelvin metre/watt.

Although techniques exist to take direct measurements of soil resistivity on site, it is not always possible to interpret these accurately, particularly as resistivity depends, to a large extent, upon soil compaction, backfilled soil not necessarily assuming the original density when returned to the cable trench.

As a general rule, in moist situations, a figure of 0.8–1.0 K m/W

Table 6 Typical soil thermal resistivities

Thermal resistivity (K m/W)	Soil conditions	Weather conditions
0.7	Very moist	Continuously moist
1.0	Moist	Regular rainfall
2.0	Dry	Seldom rains
3.0	Very dry	Little or no rain

may be adopted, whilst for clay or loam, a figure of 1.2 K m/W is a good representative figure for the UK.

Guidance in determining suitable figures for a range of soil structure is available in ERA Report 69-30 Part 1.

Where unsuitable soils are encountered, or drying out of the soil is liable to occur, it is the usual practice to surround the cable(s) with material having known thermal properties. Such material is known as 'thermal' or 'controlled' backfill.

These materials have good thermal resistivity in dry conditions and can be compacted tightly about the cable to provide good heat transfer. It is also possible to use a weak sand/shingle cement mix having a low water content for the purpose. Standard UK conditions for cables laid directly in the ground are as follows:

1. Ground temperature – 15°C
2. Soil thermal resistivity – 1.2 K m/W
3. Adjacent circuits – at least 1.8 m distance between
4. Depth of laying – 0.5 m for 1 kV cables; 0.8 m for cables above 1 kV up to 33 kV

Typical rating factors are given in Tables 7 and 8.

Reference has been made elsewhere to the need to apply a rating factor to establish current ratings for power cables drawn into ducts and, in this connection, published data are available in the

Table 7 Rating factors for various ground temperatures

Cable insulation	Maximum conductor operating temperature (°C)	Ground temperature (°C)							
		10	15	20	25	30	35	40	45
Paper	65	1.05	1.0	0.95	0.89	0.84	0.77	0.71	0.63
Paper	70	1.04	1.0	0.95	0.90	0.85	0.80	0.74	0.67
Paper	75	1.04	1.0	0.96	0.92	0.88	0.83	0.78	0.73
PVC	70	1.04	1.0	0.95	0.90	0.85	0.80	0.74	0.67
XLPE	90	1.03	1.0	0.97	0.93	0.89	0.85	0.81	0.77

Table 8 Rating factors for thermal resistivity of soil

Conductor size (mm^2)	Soil thermal resistivity (K m/W)						
	0.8	0.9	1.0	1.5	2.0	2.5	3.0
Single-core cables							
Up to 150 mm^2	1.16	1.11	1.07	0.91	0.81	0.73	0.67
From 185 mm^2 to 400 mm^2	1.17	1.12	1.07	0.90	0.80	0.72	0.66
From 500 mm^2 to 1200 mm^2	1.18	1.13	1.08	0.90	0.79	0.71	0.65
Multicore cables							
Up to 16 mm^2	1.09	1.06	1.04	0.95	0.86	0.79	0.74
From 25 mm^2 to 150 mm^2	1.14	1.10	1.07	0.93	0.84	0.76	0.70
From 185 mm^2 to 400 mm^2	1.16	1.11	1.07	0.92	0.82	0.74	0.68

IEE Regulations for Electrical Installations and also in cable manufacturers' literature.

Table 9 exemplifies the rating factors resulting when a single-way duct is laid at varying depths. By filling ducts with a mixture of bentonite, cement and sand mixed with water the heat transfer properties of the duct are considerably enhanced, the thermal resistivity properties of the mixture being less than 1.2 Km/W, resulting in an increase in rating of approximately 10%.

Cables that are installed in air, either in cleats, racks or cable tray, have ratings that are governed by the thermal resistivity of the cable materials, the ambient temperature and the proximity of other current-carrying conductors. Spacings of 75 mm to 105 mm, depending upon conductor size, should, provided adequate free ventilation exists between the cables, be sufficient to avoid derating the cable.

Where restrictions in space occur, reference should be made to ERA Report 74-27 'Heat dissipation for cables in air' and ERA Report 74-28 'Heat dissipation for cables on perforated steel trays'. Further information is available in the IEE Regulations for Electrical Installations and literature published by cable manufacturers.

The effect of grouping on cable rating is shown in Table 10.

Current rating and cable losses

Table 9 Rating factors for depth of laying – to centre of duct or trefoil group of ducts – average values. Derived from data published by BICC plc

Depth of laying (m)	600–1000 V cables		1900–3300 V to 12 700–22 000 V cables	
	Single core	Multicore	Single core	Multicore
0.50	1.00	1.00	–	–
0.60	0.98	0.99	–	–
0.80	0.95	0.97	1.00	1.00
1.00	0.93	0.96	0.98	0.99
1.25	0.90	0.95	0.95	0.97
1.50	0.89	0.94	0.93	0.96
1.75	0.88	0.94	0.92	0.95
2.00	0.87	0.93	0.90	0.94
2.50	0.86	0.92	0.89	0.93
3.00 or more	0.85	0.91	0.88	0.92

Table 10 Group rating factors for multicore cables in horizontal formation – average values. Derived from data published by BICC plc

	Number of cables in group	Spacing (centre to centre) (m)				
		Touching	0.15	0.3	0.45	0.6
600–1000 V cables	2	0.81	0.87	0.91	0.93	0.94
	3	0.70	0.78	0.84	0.87	0.90
	4	0.63	0.74	0.81	0.86	0.89
	5	0.59	0.70	0.78	0.83	0.87
	6	0.55	0.67	0.76	0.82	0.86
1900–3300 to 12 700–22 000 V cables	2	0.80	0.85	0.89	0.90	0.92
	3	0.68	0.75	0.80	0.84	0.86
	4	0.62	0.70	0.77	0.80	0.84
	5	0.57	0.66	0.73	0.78	0.81
	6	0.55	0.63	0.71	0.76	0.80

Power Cable Installation Practice

To summarize, the factors which should be taken into account when determining the rating of a power cable are as follows:

1. *Temperature.* The ambient temperature in which a cable operates is important as this can limit the maximum permissible temperature at which the cable can safely operate, and thus, in some circumstances, limit the load carried.
2. *Cable design.* In order that the cable may operate at maximum efficiency the materials used in the construction of the cable should be suitable to safely transmit the heat generated in the conductors to the outer surface of the cable.
3. *Installation conditions.* Provided that adequate spacings are maintained, cables installed in free air should have a greater measure of heat dissipation than cables laid in the ground. The rating of cables buried in the ground vary with the depth of burial.
4. *Effects of neighbouring cables.* Heat generated by nearby cables and other local heat sources have to be taken into account in installation design. The basis for the calculations involved may be obtained by reference to IEC 287.
5. *Deviation from standard conditions.* Other factors which will have a bearing on the operation of the cable need to be assessed and corrective factors applied where necessary.

Voltage drop

Generally, voltage drop only assumes importance in circuits operating below 1000 V.

Regulation 525 of the IEE Regulations for Electrical Installations (16th edn) specifies that the voltage at the terminals of a fixed current-using equipment shall comply with the relevant British Standard or, where this does not apply, the voltage shall be such as not to impair the safe working of the equipment. The Regulation (525-01-02) states that the voltage drop between the supply

Current rating and cable losses

terminals and the fixed current-using equipment should not exceed 4 per cent of the nominal supply voltage. Motor starting conditions are disregarded. The voltage drop in a circuit may be calculated from the tables provided in the IEE Regulations or from cable manufacturers' published data. These tables give values for the voltage drop of cables in units of millivolt per ampere per metre, or the impedance per unit length of the cable, Z. An additional factor which needs to be considered is the differing dissipation in single phase and three phase circuits which requires the introduction of a correction factor resulting in an effective impedance per unit length, Z_{eff} as follows:

$$\text{single-phase circuit } Z_{eff} = 2Z$$
$$\text{three-phase circuit } Z_{eff} = \sqrt{3}Z$$

In a single-phase circuit both the phase and neutral conductors form the overall impedance of the circuit, thus producing the expression $2Z$.

However, in a three-phase circuit the voltage drop will be $\sqrt{3}$ times that of one conductor so that the formula becomes $\sqrt{3}Z$.

When selecting a suitable cable for a circuit, account should be taken both of the current to be carried and also the type of protective device which is to be fitted.

Tables are provided in the IEE Regulations for Electrical Installations, and also in cable makers' literature, where values of voltage drop are tabulated for the current of one ampere for a distance of one metre of the cable route based upon the formula shown above.

Example

It is required to transmit 100 A in a three-phase circuit over a distance of 200 m, in air, by means of a three-core PVC-insulated, armoured and sheathed cable, the supply voltage being 415 V, 50 Hz.

Power Cable Installation Practice

Let V_d be the voltage drop in volts.

Then $$V_d = \frac{Z \times I \times L}{1000} \text{ or } Z = \frac{V_d \times 1000}{I \times L}$$

where I = current in amperes per phase
 L = route length in metres
 Z = approximate volt drop/ampere/metre

The maximum permissible volt drop is 4 per cent of 415 V which equals 16.6 V.

Therefore, substituting the values provided in the example, we obtain the following calculation:

$$Z = \frac{16.6 \times 1000}{100 \times 200} = 0.8 \text{ mV}$$

If we now refer to the IEE Regulations for Electrical Installations (16th edn), we find, on page 195, Table 4D4B, 'Voltage drop (per ampere per metre)' for multicore armoured PVC cables. The fourth column of the table lists the calculated voltage drop (mV) in respect of three- or four-core cable, three-phase AC. The column lists the voltage drop with respect to resistance (r) and reactance (X) as well as impedance (Z).

As it is required to limit the maximum permissible voltage drop to 4 per cent, the voltage drop $A^{-1} m^{-1}$ must be equal to or less than 0.8 mV. Consulting the table we find that, in the prescribed conditions, a 50 mm^2 three- or four-cored cable is rated at 163 A with an approximate voltage drop $A^{-1} m^{-1}$ of 0.81 mV.

If, however, it is considered that a level of 0.81 mV is too close a tolerance, the next larger size conductor will need to be selected, i.e. 70 mm^2 with a voltage drop value of 0.57 mV.

The above method of determining conductor size is generally quite satisfactory, but there can be occasions when it is necessary to carry out more precise calculations. Such circumstances arise when the actual current differs considerably from the published rating, temperature conditions affect conductor resistance, or derating factors arise as a result of site installation conditions.

Reference

McAllister, D. (1982) *Electric Cables Handbook*, Granada.

FIVE

Short-circuit performance of power cables

In recent years the expansion and increased interconnection of systems has led to high levels of energy being channelled, via the cable system, into any fault that may develop. It therefore follows that the size of a power cable conductor can be determined by its capacity to carry a heavy short-circuit current rather than the current arising from the load being supplied.

When a short-circuit occurs there is, firstly, a large inrush of current for a few cycles followed by a rapid fall. At this time the energy being delivered into the fault is expended in the form of heat in the cable conductors. Until the circuit protection operates, usually between 0.2 and 3.0 s, there is a violent increase in conductor temperature. If a cable is operating at its maximum permissible continuous rating at the time the short-circuit is initiated, the temperature rise to which the cable is subjected is the main factor in determining the cable rating.

In addition to the generation of heat the high levels of current flowing produce thermomechanical and electromagnetic forces which are proportional to the square of the current. The magnetic forces acting upon the conductors, by reason of current flow, produce disruptive stresses upon the installation and therefore impose limiting factors upon cable design, joints and terminations and upon the general design parameters of the whole installation.

If we consider the development of the short-circuit in Figure 9, which assumes that the circuit remains uninterrupted for a

Short-circuit performance of power cable

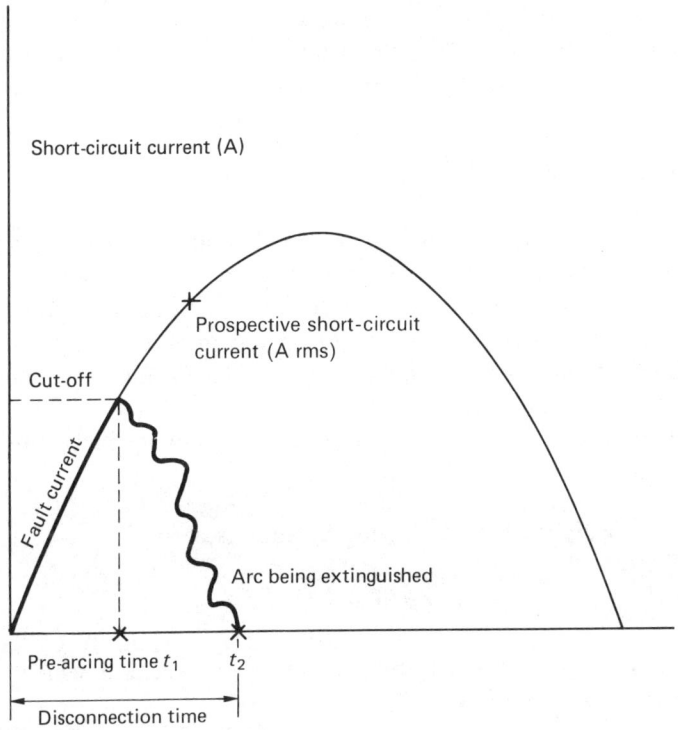

Figure 9 *Growth of a fault current*

half-cycle, there will be a steep rise towards the cut-off point when the short-circuit is interrupted and an arc is formed; this is called the 'pre-arcing' time (t_1).

When the circuit is opened the current falls to zero as the arc is being extinguished; this is the 'disconnection' time and is indicated by t_2. During the pre-arcing time the circuit-breaking device is allowing energy to flow into the circuit. This energy is defined as the *pre-arcing let-through energy* and may be shown as:

$$I_f^2 t_1 \qquad (1)$$

47

where I_f is the short-circuit current and t_1 the pre-arcing time. The total let-through energy from the initiation to the disconnection of the fault may be expressed as:

$$I_f^2 t_2 \qquad (2)$$

where t_2 is the time elapsed from the commencement of the flow of fault current to its cut-off point.

By the introduction of a factor k which takes into account the type of conductor metal and insulating material, the expression $k^2 s^2$ is derived, s being the cross-sectional area of the conductor.

This, then, is the limiting factor for the amount of heat energy that a cable can withstand. Therefore, the let-through heat energy should not exceed $k^2 s^2$.

From Equation 2 we can develop the following expression:

$$I_f^2 t_2 = k^2 s^2$$

Transposing for t the following formula is derived:

$$t = \frac{k^2 s^2}{I_f^2}$$

t being the maximum disconnection time in seconds which will preserve cable integrity.

The above formula is an approximation and is, together with k for a number of combinations of aluminium or copper conductors insulated with a range of materials, fully detailed in Regulation 434-03-03 of the IEE Regulations.

The formula may be adapted for general use as follows:

$$\text{Short-circuit current} = k \times S T^{-\frac{1}{2}}$$

where S = cross-sectional area of conductors in mm^2
T = duration of the short-circuit in seconds

Table 11 may be used for short-circuit calculations; the parameters are based upon the assumption that the cable is functioning at maximum permissible operating temperature, on a continuous basis, at the time the short-circuit occurs.

Table 11 Short-circuit currents for standard cables

Cable insulation/type	Conductor metal	Temperature rise (°C)	Short-circuit current (A)
Paper			
1–6 kV belted	Copper	80–160	$108 \times ST^{-\frac{1}{2}}$
	Aluminium	80–160	$71 \times ST^{-\frac{1}{2}}$
10–15 kV belted	Copper	65–160	$119 \times ST^{-\frac{1}{2}}$
	Aluminium	65–160	$78 \times ST^{-\frac{1}{2}}$
10–15 kV screened	Copper	70–160	$115 \times ST^{-\frac{1}{2}}$
	Aluminium	70–160	$76 \times ST^{-\frac{1}{2}}$
20–30 kV screened	Copper	65–160	$119 \times ST^{-\frac{1}{2}}$
	Aluminium	65–160	$78 \times ST^{-\frac{1}{2}}$
PVC			
⩽300 mm² (1–3 kV)	Copper	70–160	$115 \times ST^{-\frac{1}{2}}$
	Aluminium	70–160	$76 \times ST^{-\frac{1}{2}}$
>300 mm²	Copper	70–140	$103 \times ST^{-\frac{1}{2}}$
	Aluminium	70–140	$68 \times ST^{-\frac{1}{2}}$
XLPE and EPR	Copper	90–250	$143 \times ST^{-\frac{1}{2}}$
	Aluminium	90–250	$94 \times ST^{-\frac{1}{2}}$

S = area of conductor (mm²)
T = duration of short-circuit (s)

Furthermore, the figures derived are based upon certain limiting factors: paper cables are limited to 160°C, on the assumption that conductor joints are soldered, and XLPE, together with EPR, have compression-type mechanical joints which may be operated up to 250°C.

Table 12 provides details of the temperature limits which apply to the various materials used in power cable construction and jointing which need to be taken into consideration under short-circuit conditions. It should be noted, particularly, that figures for

Table 12 Short-circuit temperature limits

Material or component	Temperature (°C)
Material	
Paper insulation	250
PVC insulation $\leqslant 300\,\text{mm}^2$	160
PVC insulation $> 300\,\text{mm}^2$	140
PVC insulation $\geqslant 6.6\,\text{kV}$	140
PVC oversheath	200
Natural rubber	200
Butyl rubber	220
Polyethylene	150
XLPE and EPR	250
Silicone rubber	350
CSP oversheath	220
Components	
Soldered conductor joints	160
Compression joints	250
Lead sheaths – unalloyed	170
Lead sheaths – alloyed	200

oversheaths have been included, as, being in contact with the armour wires, the material can be subjected to higher temperatures than in the case of an unarmoured power cable.

The data given in Table 11 are expressed in graphical form for paper, PVC and XLPE in Figures 10, 11 and 12.

Thus far we have been concerned with symmetrical three-phase faults, that is to say when the short-circuit has taken place between current-carrying conductors. In the case of an earth fault, which is asymmetrical, it is necessary to take account of the fact that, in a paper cable, the fault current will be carried by the armour and lead sheath whilst in a PVC- or XLPE-insulated cable it will be carried by the armour alone.

It will be seen from Table 12 that a lead sheath will be seriously

Short-circuit performance of power cable

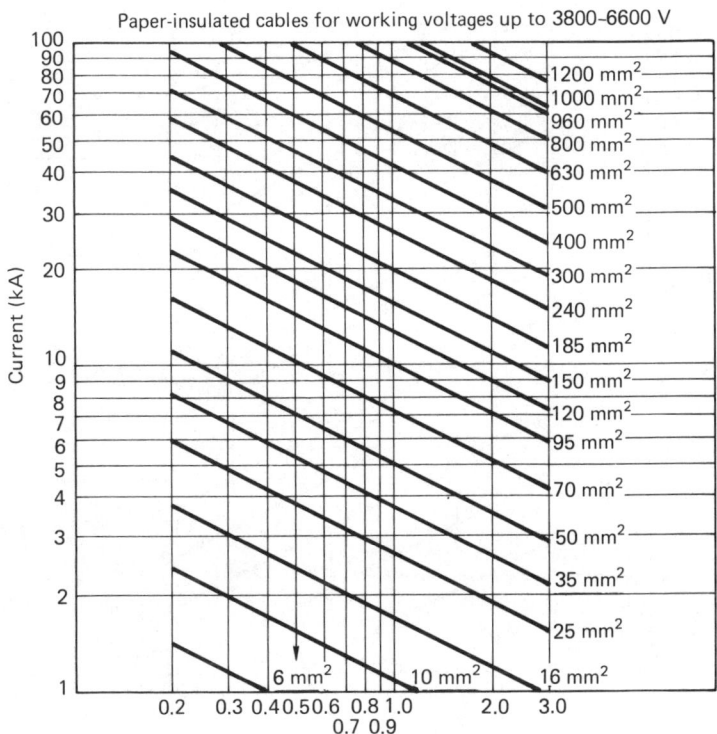

Figure 10 *Short-circuit ratings for paper. Copper conductors, paper insulated, lead sheathed, single wire armoured, PVC sheath. The values of fault current shown in the graph are based upon the cable being fully loaded at the initiation of the short-circuit with a conductor temperature of 80°C up to 6.6 kV and 65°C for 11 kV belted cable. The conductor temperature at the end of the short-circuit is 160°C. A correction factor of 1.07 should be applied for single and three-core screened cables in the range 6350–11 000 V and 8700–15 000 V. A correction factor of 1.10 should be applied for three-core belted 6350–11 000 V cables and all 12 700–22 000 V and 19 000–33 000 V cables*

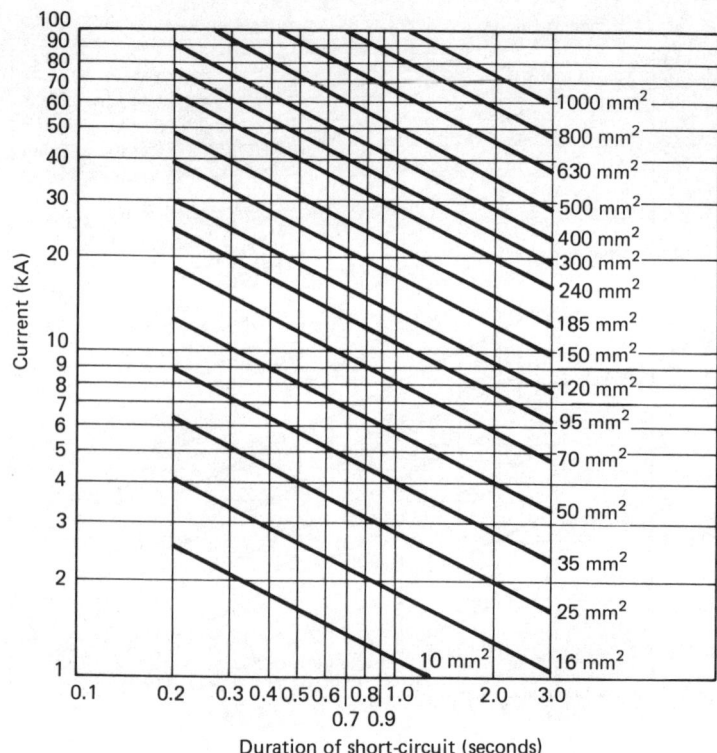

Figure 11 *Short-circuit ratings for PVC. Copper conductors, PVC insulated, armoured, PVC sheathed, to BS 6346. The values of fault current shown in the graph are based upon the cable being fully loaded at the initiation of the short-circuit with a conductor temperature of 70°C and a final conductor temperature of 160°C for conductor sizes $\leqslant 300\,mm^2$ and 140°C for conductor sizes $> 300\,mm^2$*

damaged if subjected to excessive temperatures and PVC oversheaths can be similarly damaged by high levels of armour wire temperature. Similarly, soldered joints in conductors are adversely affected by the increased conductor heat under short-circuit conditions. Tables 13 to 16 give details of the maximum allowable

Short-circuit performance of power cable

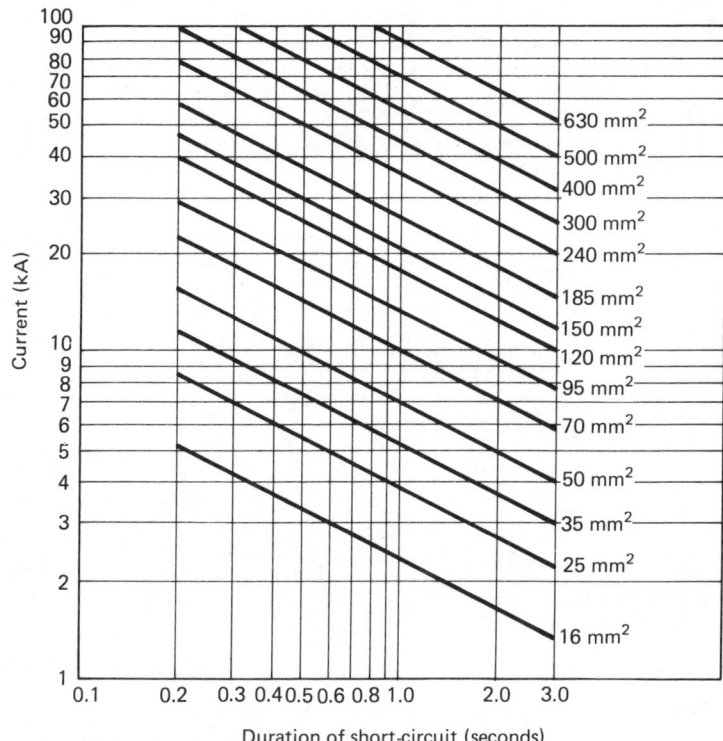

Figure 12 *Short-circuit ratings for XLPE. Copper conductors, XLPE insulated, single wire armoured, PVC sheathed. The values of fault current shown in the graph are based upon the cable being fully loaded at the initiation of the short-circuit with a conductor temperature of 90°C and a final conductor temperature of 250°C. It should be noted that accessories employed must be compatible*

asymmetrical fault currents which may be tolerated by power cables in general use. The figures are based upon stranded conductor cables which have been subjected to fault conditions for 1 s. To compute values for other periods of time the values should be divided by the square root of the time in seconds.

Power Cable Installation Practice

Table 13 Maximum allowable asymmetrical current to earth (1 s rating) for single-core PILS cables (BICC plc)

Conductor size (mm^2)	0.6– 1 kV (kA)	1.9– 3.3 kV (kA)	3.3– 6.6 kV (kA)	6.35– 11 kV (kA)	8.7– 15 kV (kA)	12.7– 22 kV (kA)	19– 33 kV (kA)
50	1.4	1.6	1.7	1.9	2.3	2.9	3.8
70	1.6	1.8	1.9	2.3	2.5	3.1	4.0
95	1.9	2.0	2.3	2.5	2.9	3.6	4.6
120	2.2	2.4	2.5	2.9	3.2	3.8	4.8
150	2.5	2.6	3.0	3.1	3.6	4.3	5.3
185	2.9	3.0	3.2	3.6	3.9	4.6	5.8
240	3.3	3.4	3.8	4.0	4.6	5.3	6.3
300	3.9	4.0	4.2	4.7	4.9	5.7	6.8
400	4.7	4.7	4.9	5.5	5.7	6.6	7.7
500	5.6	5.6	5.7	6.1	6.8	7.6	8.8
630	6.6	6.6	6.8	7.2	7.9	8.8	10.1
800	7.7	7.7	7.9	8.3	9.1	10.1	11.4
1000	9.0	9.0	9.2	10.2	10.5	11.6	13.0
960*				9.8	10.6	11.7	13.2
1200*				11.8	12.2	13.4	14.9

* Milliken-type conductors – copper only.

It is assumed that in the case of paper-insulated, lead-sheathed and single wire armoured cables the fault current is carried by both the sheath and the armour.

Under short-circuit conditions a multicore cable is subjected to electromagnetic forces which repel the cores from one another and thus tend to rupture the material binding the cores together. This problem can become particularly acute in the case of unarmoured paper-insulated cables as the insulation can be damaged by the outward movement of the cores.

With conductor sizes in excess of 185 mm^2 and current levels exceeding 30 kA under short-circuit conditions it is necessary to limit the ratings which have been calculated from the formula $kST^{-\frac{1}{2}}$ and these levels for unarmoured cables are given in Table 17.

Table 14 Maximum allowable asymmetrical fault current to earth (1 s rating) for multicore PILS/SWA cables (BICC plc)

Conductor size (mm²)	0.6–1 kV Three core (kA)	0.6–1 kV Four core (kA)	1.9–3.3 kV Three core (kA)	3.8–6.6 kV Three core (kA)	6.35–11 kV Three core (belted) (kA)	6.35–11 kV Three core (screened) (kA)	8.7–15 kV Three core (screened) (kA)	12.7–22 kV Three core (screened) (kA)	19–33 kV Three core (screened) (kA)
4	3.1	3.4							
6	3.4	3.6							
10	3.8	4.9							
16	4.4	5.0	5.0	6.4					
25	5.2	6.1	5.1	7.3	10.1	10.1			
35	6.0	6.7	6.0	8.7	11.3	11.3			
50	6.8	8.9	6.6	9.6	11.3	10.9	12.8		
70	8.7	10.3	8.5	10.3	14.3	12.0	16.2	17.9	
95	10.1	12.0	9.9	11.6	15.7	15.4	16.4	19.5	
120	11.4	15.4	11.3	15.0	17.5	17.1	17.8	20.6	26.9
150	15.3	17.9	14.4	16.5	18.6	18.7	20.1	21.3	27.5
185	17.1	20.1	16.0	17.9	20.2	20.2	21.3	23.2	33.5
240	19.4	23.6	17.5	19.8	22.1	22.1	22.9	24.9	35.7
300	21.8	26.3	19.8	22.6	25.0	24.6	24.9	26.7	37.0
400	25.0	34.5	22.6	25.0	31.6	31.6	32.2	32.2	39.5
			29.8	32.6	35.6	35.7	35.1	36.8	43.0
							39.3	39.9	46.3
								44.4	51.0

Table 15 Maximum allowable asymmetrical fault current to earth (1 s rating) for PVC-insulated wire armoured cables with copper conductors (BICC plc)

Conductor size (mm^2)	Aluminium armour		Steel wire armour				1.9–3.3 kV Three-core (kA)
	0.6–1 kV Single core (kA)	1.9–3.3 kV Single core (kA)	0.6–1 kV				
			Two-core (kA)	Three-core (kA)	Four-core equal (kA)	Four-core reduced neut. (kA)	
1.5			0.7	0.7	0.7		
2.5			0.8	0.8	0.9		
4.0			0.9	1.0	1.5		
6.0			1.0	1.5	1.7		
10.0			1.8	1.9	2.1		
16.0			1.7	1.9	2.7		
25.0			2.7	2.9	3.4	3.4	3.3
35.0			2.9	3.3	3.7	3.6	3.6
50.0	3.1	3.5	3.3	3.7	5.4	4.2	4.0
70.0	3.5	3.9	3.7	5.3	6.1	5.9	5.4
95.0	4.0	5.7	5.4	6.1	7.0	6.9	6.1
120	5.7	6.2	5.8	6.6	9.7	9.5	6.6
150	6.4	6.5	6.4	9.3	10.8	10.4	9.1
185	7.0	7.0	8.9	10.2	11.7	11.4	9.7
240	7.8	7.8	9.9	11.4	13.2	12.7	10.4
300	8.6	8.6	11.0	12.7	14.7	14.3*	11.4
						14.7†	12.7
400	12.2	12.2	12.3	14.0	20.6	19.9	14.0
500	13.4	13.4					
630	14.6	14.6					
800	20.6	20.6					
1000	22.9	22.9					

* 300/150 mm^2

Table 16 Maximum allowable asymmetrical fault current to earth (1 s rating) for XLPE-insulated wire armoured cables with copper conductors

Conductor size (mm^2)	Aluminium armour		Steel wire armour				1.9–3.3 kV Three-core (kA)
	0.6–1 kV Single core (kA)	1.9–3.3 kV Single core (kA)	0.6–1 kV				
			Two-core (kA)	Three-core (kA)	Four-core equal (kA)	Four-core reduced neut. (kA)	
16			1.7	1.7	1.9		3.1
25			1.7	2.4	2.7	2.7	3.1
35			2.4	2.7	3.1	3.0	3.3
50	1.8	2.7	2.6	3.0	3.5	3.3	4.6
70	2.7	3.1	3.1	3.5	5.1	5.0	5.1
95	3.1	3.3	4.4	5.0	5.7	5.6	5.7
120	3.3	4.8	4.9	5.5	8.0	6.3	7.8
150	4.8	5.1	5.4	7.8	9.0	8.6	8.4
185	5.4	5.7	7.4	8.6	9.9	9.7	9.0
240	6.0	6.0	8.4	9.7	11.3	10.9	9.9
300	6.4	6.8	9.2	10.5	12.4	11.8*	10.9
400	9.1	9.1				12.4†	
500	10.5	10.5					
600	11.8	11.8					

* 300/150 mm^2
† 300/185 mm^2

Power Cable Installation Practice

Table 17 Current limitation, due to bursting under short-circuit conditions, of unarmoured belted multicore paper-insulated cables (BICC plc)

Voltage (kV)	Aluminium conductor		Copper conductor	
	Size (mm^2)	Short-circuit rating (kA)	Size (mm^2)	Short-circuit rating (kA)
0.6–1	240	33	120	25
	300	35	150	27
	400	37	185	29
			240	33
			300	36
			400	38
1.9–3.3 and 3.8–6.6	240	33	185	33
	300	35	240	35
	400	38	300	37
			400	38
6.35–11	240	39	185	36
	300	41	240	39
	400	43	300	41
			400	43

Armouring wires provide adequate reinforcement to restrict the bursting forces from rupturing the cable. Similarly, where metallic screening has been applied there is a restriction on movement of the cores.

Polymeric insulation, either of the thermoplastic or the thermosetting type, can withstand short-circuit damage to a greater extent than paper insulation but, if unarmoured, short-circuit current limitation against bursting would need to be applied.

It should also be noted that single-core cables should be installed with adequate cleats which will restrict the movement of the conductors under short-circuit conditions.

Considerable heat is generated under short-circuit conditions which produces a number of undesirable effects on the cable installation. The problem of soldered joints has already been mentioned but, in addition, the expansion of the conductors, creating longitudinal thrust and excessive movement of the cables, if not adequately secured, is a significant danger area although single-core cables, by reason of construction, are not so subject to these difficulties.

Joints are a particular problem by reason of the restriction of the surrounding ground on the outer surfaces of the cable and movement of the cores within the cable and joint itself.

Soldered joints are limited to 160°C and compression or mechanical connectors need to be stable at elevated temperatures to maintain a low-resistance connection. Similarly, the enclosure forming the joint container should be of suitable construction to accommodate the high fluidic pressure which will be generated by the expansion of the resin-based impregnant used in paper cables, which can then flow into the body of the accessory, softening the compound filling.

It is important to note that cables installed in air should be adequately spaced in respect of supports and rigidly cleated where necessary to prevent excessive generation of either expansive or compressive forces in proximity to accessories.

Thermoplastic insulation and oversheaths can be deformed by excessive pressures applied over small areas by excessive bending radii and inappropriate clamping devices.

The larger sizes in the EPR/XLPE categories need to be treated with similar caution in view of the higher temperature of 250°C for these materials.

Calculation of short-circuit current

Whilst it is not the remit of this book to deal with the finer points of calculating prospective values of short-circuit currents at various circuit levels, nevertheless IEE Regulation 434-02-01 states: 'The prospective fault current, under both short-circuit and earth fault conditions, at every relevant point of the complete installation

Power Cable Installation Practice

shall be determined. This shall be done by either calculation or measurement."

It is not always possible to obtain the necessary data, when designing an electrical installation, to accurately calculate the precise value of the prospective short-circuit current, and quite often a rough estimate will suffice. This can be obtained in the following manner.

If a supply voltage of 415 V falls to 410 V with a load of 50 A connected, then the approximate short-circuit current (I_{sc}) is:

$$z = \frac{\text{Circuit volt drop}}{\text{Load current}} = \frac{5\,\text{V}}{50\,\text{A}} = 0.1\,\Omega$$

Therefore:

$$I_{sc} = \frac{V}{z} = \frac{415}{0.1} = 4150\,\text{A}$$

The value of fault current which flows in any system in short-circuit conditions is limited only by the impedances of the system. Therefore, the basis of short-circuit calculation for any system must be the evaluation of these impedances.

Whilst impedance is the vectorial value of resistance and reactance, the value of resistance is relatively small in the case of machines and transformers so that it is usual to express the reactance of these items in percentage terms. That is to say that an item which has a reactance of x per cent will have a reactive voltage drop of x per cent of the normal voltage when carrying full load current corresponding to the normal rating.

In the case of power cables, values of resistance and reactance are obtainable from the manufacturers based upon ohms per kilometre of cable and can be used to provide quite precise levels of impedance, thus enabling more accurate values of prospective fault current to be established.

Most practical calculations can be effected by means of the 'percentage reactance method', which makes use of the reactance values provided by the equipment manufacturers.

Short-circuit performance of power cable

Initially, a base figure needs to be established; this can be the value of the largest piece of plant connected, or the total value of all the plant on the system, or alternatively it can be an arbitrary figure, say 100 000 kVA. Therefore, a piece of plant rated at 10 000 kVA, and a reactance of 10%, will have 100% reactance on the 100 000 kVA base.

Using the following expression, it is possible to bring all reactance percentages to the common base:

$$\text{Base reactance per cent } (X_b) = \frac{\text{Base kVA}}{\text{Plant kVA}} \times \text{Plant reactance per cent } (X_p) \quad (3)$$

For example, in the case of the 10 000 kVA item above,

$$X_b \text{ per cent} = \frac{100\,000 \times 10}{10\,000} = 100 \text{ per cent}$$

Having obtained the ohmic values of resistance and reactance of the cable involved, the percentage reactance may be expressed as follows:

Percentage reactance at given kVA base for ohmic value X

$$= \frac{100\,000 \times \text{kVA base} \times X \text{ ohms}}{\text{Voltage}^2} \quad (4)$$

The above formulae may be applied in a relatively simple example which demonstrates the percentage reactance method of calculation.

Consider the simple system shown in Figure 13. A 2000 kVA generator is feeding into 11 kV busbars which in turn feed a remote substation via a cable link. A fault has developed on the busbars of the substation into which energy is being fed via the power cable which is 2 km in length. As only a single generator is involved there

Power Cable Installation Practice

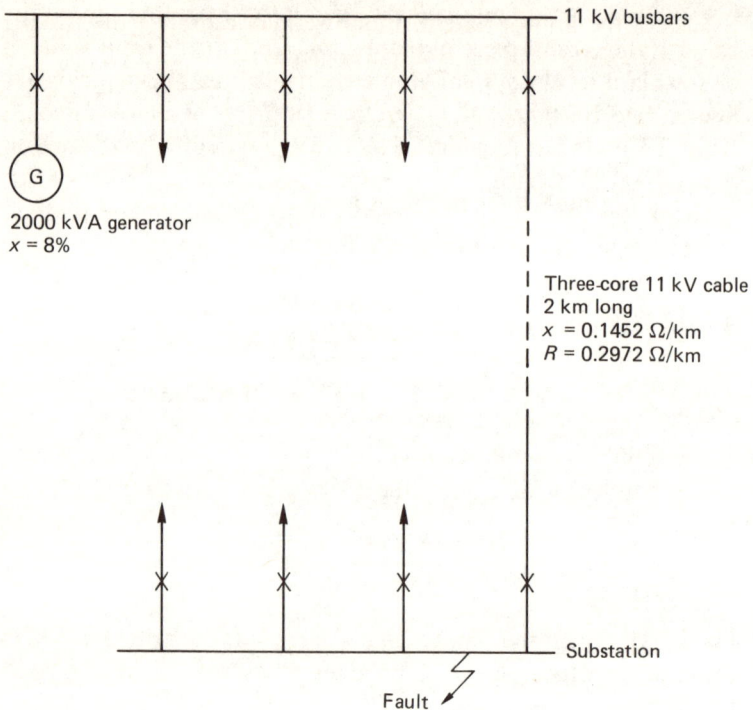

Figure 13

is no requirement to establish a common base; the instantaneous MVA may be calculated by the formula

$$\mathrm{MVA_I} = \frac{\mathrm{kVA\ Base} \times 100}{\mathrm{Percentage\ reactance} \times 1000} \qquad (5)$$

Substituting the values in the example we have:

$$\mathrm{MVA_I} = \frac{2000 \times 100}{8 \times 1000} = 25\,\mathrm{MVA}$$

By use of the appropriate formula we may establish the instantaneous symmetrical short-circuit current:

$$I_{sc} = \frac{MVA_I \times 1000}{\text{Line volts (kV)} \times \sqrt{3}} \qquad (6)$$

Substituting the values in the example:

$$I_{sc} = \frac{25 \times 1000}{11 \times 1.73} = 1313 \text{ A (rms)}$$

If we now consider the circuit it will be seen that the reactance of the generator (8%) is in series with the reactance of cable (0.2904 Ω). By applying the formula defined in Equation 4 we develop the following equation:

Percentage reactance of the cable on a 2000 kVA base

$$= \frac{100\,000 \times 2000 \times 0.2904}{11\,000 \times 11\,000}$$

$$= 0.480 \text{ per cent}$$

The percentage reactance of the circuit is found by adding the generator figure (8 per cent) to the above, making a total of 8.48 per cent. Substituting this value in Equation 5 we obtain:

$$MVA_I = \frac{2000 \times 100}{8.48 \times 1000} = 23.58 \text{ MVA}$$

The instantaneous short-circuit current is derived from Equation 6 as follows:

$$I_{sc} = \frac{23.58 \times 1000}{11 \times 1.732} = 1237 \text{ A (rms)}$$

It will be noted from the example that the total resistance of the cable is nearly twice the value of the reactance and will, therefore, make a contribution to reducing the level of the fault current.

Once again using Equation 4 we get:

Percentage resistance of the cable on a 2000 kVA base

$$= \frac{100\,000 \times 2000 \times 0.5944}{11\,000 \times 11\,000} = 0.982 \text{ per cent}$$

In order to establish the impedance of the circuit, vectorial addition must be used, impedance being derived from the following expression:

$$\text{Impedance} = \sqrt{(\text{Resistance}^2 + \text{Reactance}^2)}$$

i.e.

$$Z = \sqrt{(R^2 + X^2)}$$

Using j notation we can establish the following expressions in respect of the generator and the cable:

$$\text{Generator} = O + jX_1$$
$$\text{Cable} = R + jX_2$$
$$\text{Impedance} = R + j(X_1 + X_2)$$

Returning to our example and substituting the numerical values we have:

$$\text{Generator} = O + j8$$
$$\text{Cable} = 0.98 + j0.48$$
$$Z = 0.98 + j8.48$$
$$= \sqrt{(0.98^2 + 8.48^2)}$$
$$= 8.536 \text{ per cent}$$

Now,

$$\text{MVA}_I = \frac{2000 \times 100}{8.536 \times 1000} = 23.430 \text{ MVA}$$

Therefore the instantaneous symmetrical short-circuit current is:

$$I_{sc} = \frac{23.430 \times 1000}{11 \times 1.732} = 1229 \text{ A (rms)}$$

It will thus be seen that by taking the cable resistance into account only a small reduction in fault current results, something in the order of 1 per cent. In practice it will be found where the cable resistance is less than one third of the reactance of the *whole* circuit an error of less than 5 per cent will arise from its omission.

The calculation of asymmetrical short-circuit current is more involved in that it depends upon more detailed consideration of the circuit elements, the subject being, as previously stated, beyond the scope of this book.

SIX

Power cable accessories

The need to develop a range of accessories which would enable lengths of power cable to be connected together and also to provide a means of attachment of the power cable to electrical equipment very quickly arose as the use of electricity spread during the latter part of the nineteenth century.

Once again, Sebastian Ferranti was to the fore, having developed his paper-insulated tubular main for the Deptford project in 1888. The tubular main consisted of 20 ft (6.09 m) lengths of iron pipe which protected the wax-impregnated-paper-insulated copper tubes which were the conductors of the 10 kV system.

The total length of the Ferranti tubular main was about 27 miles, comprising four feeders which gave rise to a requirement of some 7000 straight joints together with terminations at each location where the main was connected to the equipment it served.

With the growth of the electricity supply industry the impregnated-paper-insulated cable laid in the ground, suitably protected with a lead sheath and provided with either tape or steel wire armour for mechanical protection, produced a requirement for suitable enclosures which could be buried in the ground and in which the cable could be connected to other circuits and also be sealed against the ingress of moisture which would have had disastrous effects on the paper insulation.

Similarly, there was the need to provide a suitable enclosure which would again seal the cable against moisture and still enable the circuit to be connected to switchgear, transformers and other items of plant – the so-called *dividing box*.

It was also necessary to have disconnecting points in large underground networks for convenience in isolating certain sections in the event of faults, or for testing purposes. These disconnecting boxes were sometimes set into pits in the ground or frequently in overground steel cabinets containing fuses or links controlling several feeders.

Along the cable route individual lengths of cables were connected to one another through straight-through joints, premises were serviced from the cables laid in the streets and footpaths by means of 'T' joints which, as the name implies, provide a direct connection between the consumer and the distributor cable laid outside the premises.

Very little change in the fundamental design of joint boxes took place between this latter part of the nineteenth century and the 1960s. Essentially there was a need to, firstly, prevent the ingress of moisture into the paper insulation of the cable, and secondly, provide mechanical protection to the cable at the position of the joint.

In the case of low-voltage cables, that is to say distributor and service cables, the compound sealed box was in general use (Figures 14 and 15). These boxes consisted of two halves, usually of cast iron, which, after completion of the joint, were bolted together and filled with a bituminous compound. The lead sheaths of the cables to be jointed were terminated inside the box and bonded by means of lead tape which was wrapped around the lead sheath, and also the armour, before being clamped to the body of the box. Improved electrical continuity of the sheath to the box body could be obtained by soldering a tinned copper strip to the sheath and attaching this to the box body.

The conductors were jointed using ferrules, in the case of straight-through joints, and copper claw connectors for the T-joints. After the electrical connections were made the cores

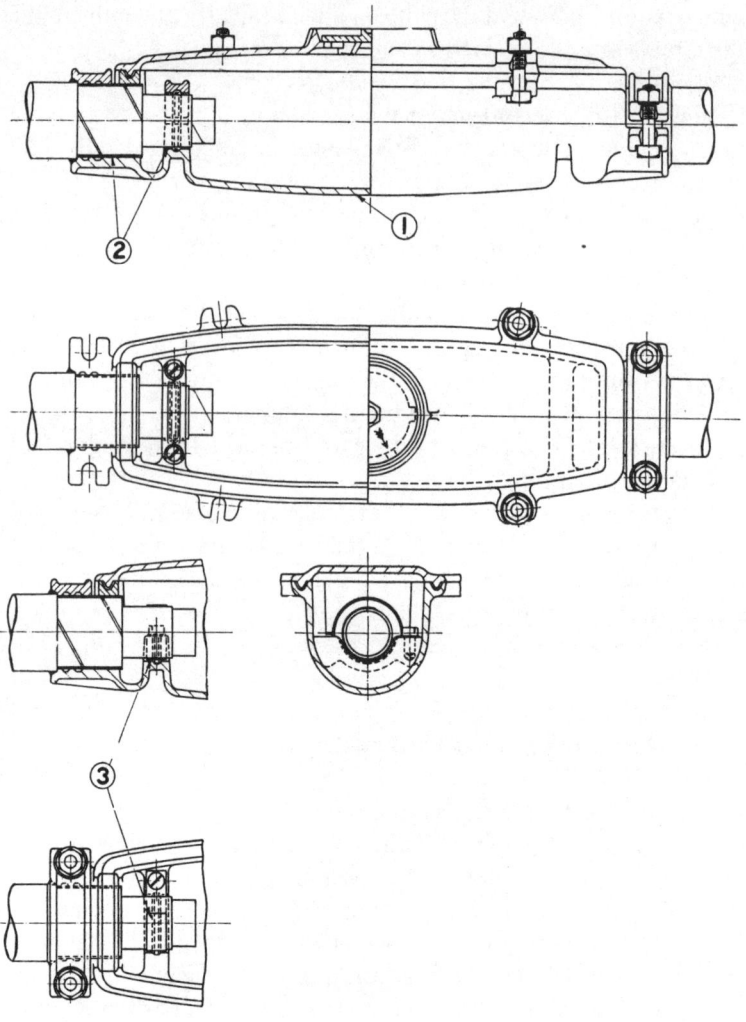

Figure 14 *Compound sealed straight joint box:* 1, cast iron box; 2, lead tape bonding for lead and armour; 3, copper strip bonding for lead

Power cable accessories

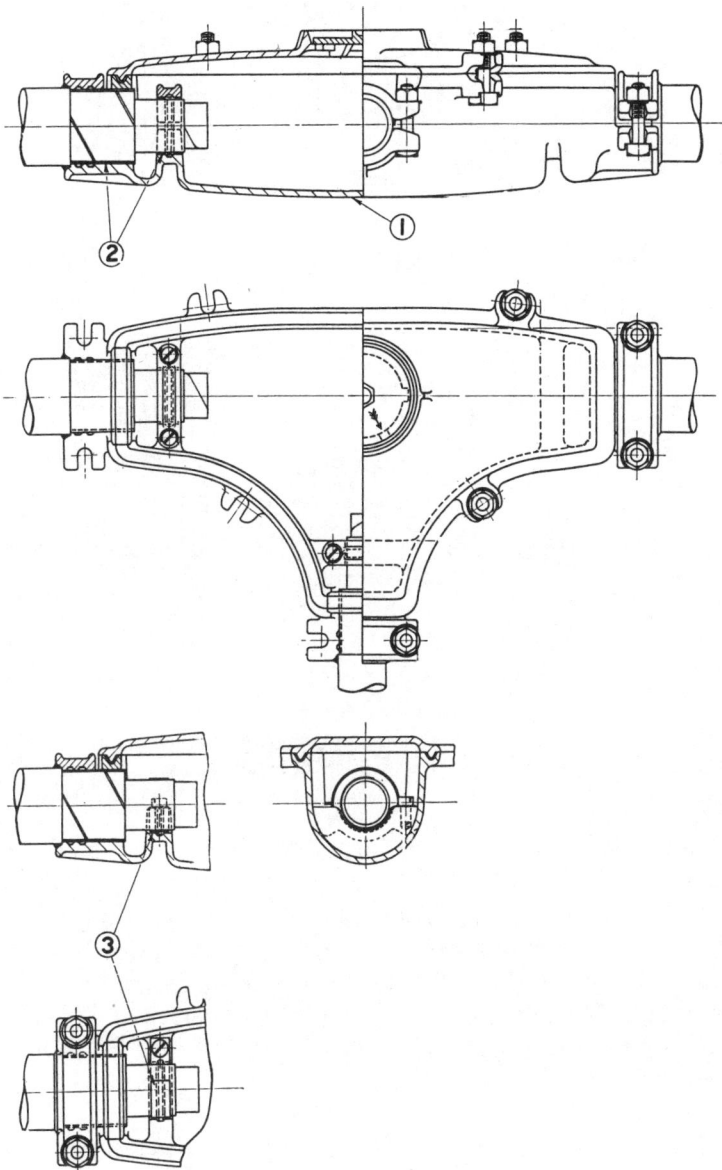

Figure 15 *Compound sealed T-box: key as Figure 14*

Power Cable Installation Practice

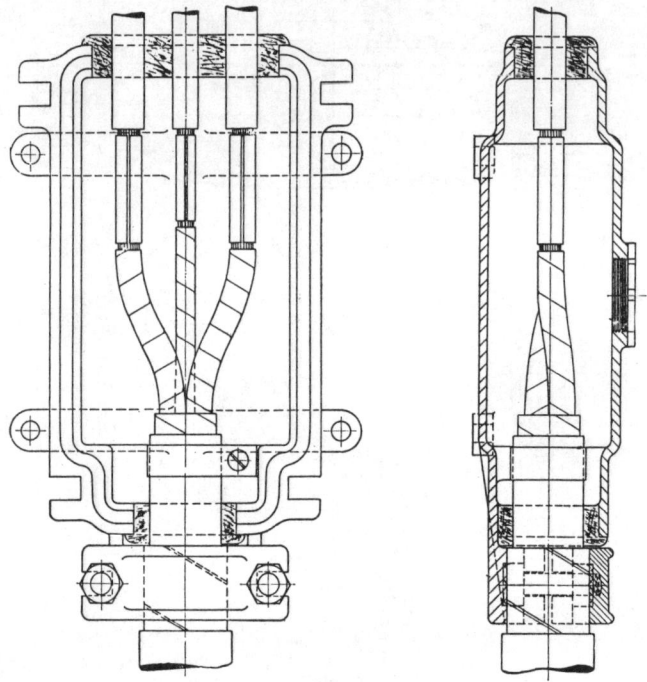

Figure 16 *Overground terminal box, low voltage: indoor type for L T cables ⩽660 V, vertical or horizontal fixing, V IR tails out*

would be taped with pre-impregnated tape and the two halves of the box united and filled with compound.

The compound sealed joint was cheap and generally quite efficient, if properly constructed, at low voltage. However, in the case of higher voltages, that is above 1000 V, it was the usual practice to fit a lead sleeve inside the outer protection box. The jointing process would be generally the same except that the lead sleeve would be attached to the lead sheaths of the cable being jointed by solder, i.e. the process of *wiping*.

The compounds used to fill the joints were graded according to

Power cable accessories

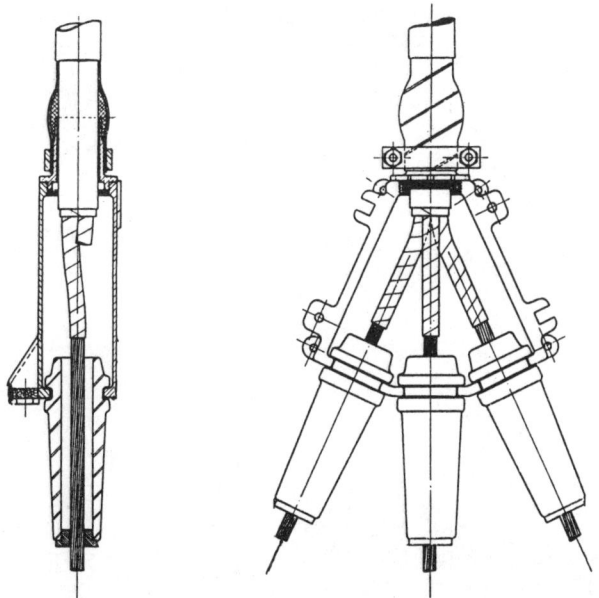

Figure 17 *Overground terminal box, high voltage: indoor type, vertical fixing, for three-core cables $\leqslant 11\,000\,V$, uninsulated outgoing conductors*

service and are dealt with more specifically in another part of this book.

Overground terminations of power cables during this period were designed to meet similar criteria except that external connections were made through VIR tails or brass studding extending through insulators (Figures 16, 17 and 18).

The use of plastic materials for power cable insulation brought into question the continued requirement for joints made in the manner described, which were expensive both in material and in manpower costs. The development of conductor connection by cold compression methods and the availability of hard setting resins as filling agents quickly led to a completely new approach to power cable joints in the 1960s and 1970s.

Power Cable Installation Practice

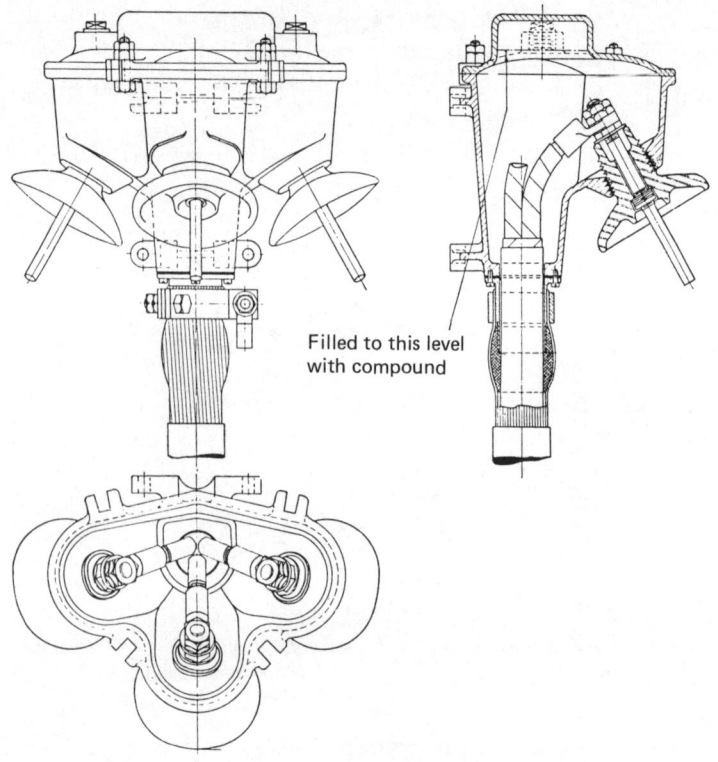

Figure 18 *Overground terminal box, high voltage: suitable for HT cables $\leq 0.3\,in^2$ $(1.85\,mm^2)$*

Further, more recent developments in moulded high-voltage cable accessories and the use of heat 'shrink-on' components has provided a very wide range of jointing methods available to the power cable installation designer.

Jointing systems involving the use of lead sleeves, solder and bituminous compound require heat, and the production of a range of accessories which would enable the jointing process to take place without the need to introduce heat allows jointing practice

Power cable accessories

to be reduced to its simplest form. The introduction of such a system was first envisaged in the 1950s but, at that time, the polyester resins used for filling the joint boxes of the period proved to be unreliable in service.

During the late 1960s through the 1970s into the early 1980s a number of resins appeared which were suitable for electrical purposes; these included polyurethane, and epoxy and acrylic resins. These products were cold-pouring and consisting of two

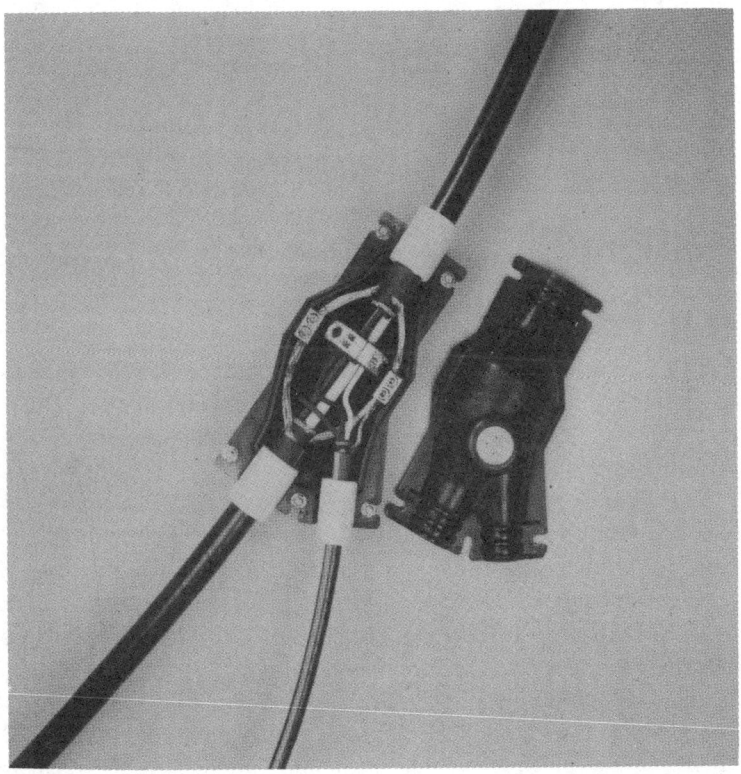

Figure 19 *Cast resin branch type joint before filling with resin (Cellpack (UK) Ltd)*

Power Cable Installation Practice

Figure 20 *Pouring resin into a parallel branch joint (Cellpack (UK) Ltd)*

parts, the resin itself and a hardening agent. When the two materials are mixed together the curing process commences, and it continues after being poured into the joint enclosure where final curing takes place.

The resin-encapsulated joint has high mechanical strength and therefore only requires a relatively light plastic shell to hold the resin in contact with the cable joint during the curing process.

Resins have been developed having special characteristics, such as the provision of low viscosity, improved tear strength and flame retardance properties, and are in general use at operating voltages up to and including 12 kV. Various configurations of the pre-shaped plastic moulded body shell, which forms the enclosure surrounding the actual cable joint and into which the resin is poured, are available, and provide straight-through, T or branch joint shapes. The body shells are supplied in two parts and are

Power cable accessories

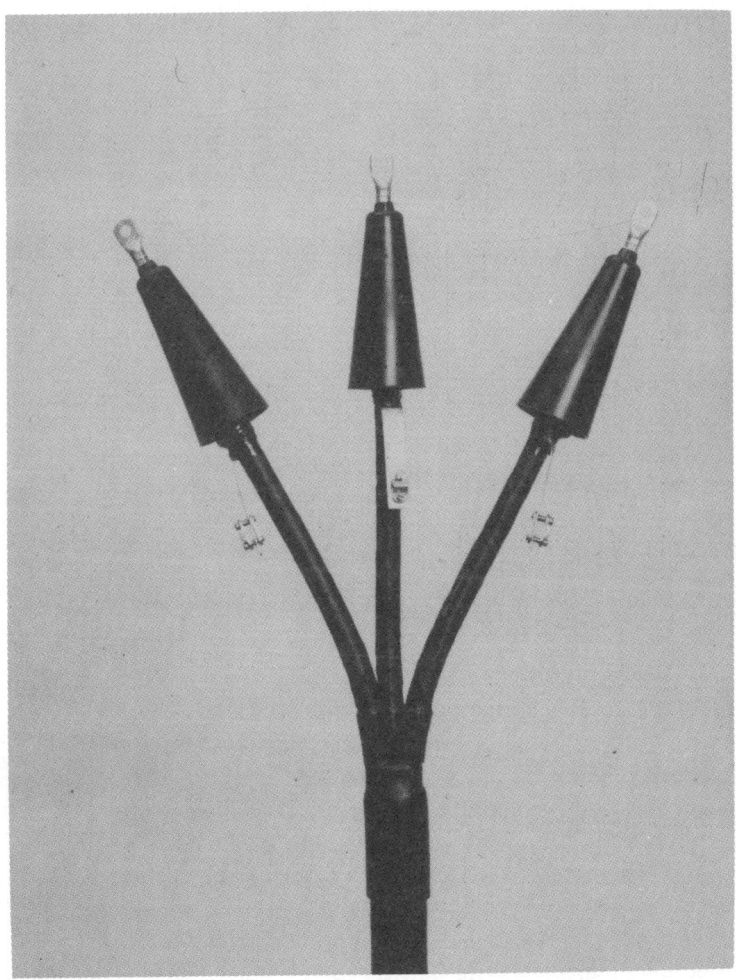

Figure 21 Slip-on type termination (Cellpack Ltd, Electrical Products Division)

Power Cable Installation Practice

Figure 22 *Slip-on terminations used for transition from overhead line to three-core plastic-insulated cable (Cellpack (UK) Ltd)*

clipped together to form the enclosure into which the resin is introduced.

More recently, the use of pre-moulded accessories has become popular in the UK. These cable terminations and joints are manufactured from elastomeric material and, after cable preparation and the application of silicone grease, the moulding slides into place on the cable. They are often referred to as 'slip-on terminations'.

The elasticity of the elastomer provides a high contact pressure between the component and the cable sheath and can also have stress control, in the form of semiconducting material, incorporated into the moulded termination.

In view of the ease with which these accessories may be installed on site, with consequent saving in time and labour costs, there is no doubt that usage will continue to increase in the future as these products are particularly suitable for XLPE-, EPR- and PVC-insulated cables and can be provided for operating voltages up to and including 33 kV.

SEVEN
Power cable jointing

I first encountered the power cable jointing operation as a young student apprentice in the pre-nationalization days of the electricity supply industry; at that time paper-insulated, lead-sheathed cables ruled supreme and good plumber-jointers were regarded as the highest exponents of individual craft skills.

Indeed, however well a cable joint may be designed, poor-quality workmanship can entirely nullify the best efforts of the designer. It is true that today with the advent of new materials, tools and improved techniques the end result is not so dependent upon the individual workmanship of the jointer but, nevertheless, lack of care in cleanliness and the careless use of tools and solvents can impair the satisfactory completion and subsequent service life of the joint and therefore the efficiency of the circuit of which the joint forms a part.

It follows, therefore, that the selection and training of jointers is of considerable importance and I would particularly stress the need to ensure that the jointer is kept up to date with new developments in order that he can identify and apply appropriate techniques to materials not previously encountered in his practical experience.

Paper-insulated lead-sheathed cables

The method of preparing a PILC cable for jointing commences with the process of removing the sheathing or serving from the cable and exposing the armouring and/or lead sheath which may be then cleansed with a suitable solvent.

Having determined the point at which the lead sheath should be removed (this will be a short distance within the joint enclosure) and also appropriately located the armouring for eventual clamping via the armour clamp, the next operation, after removing the lead sheath, is to strip off the belting papers together with the fillers etc., and prepare the conductors for terminating or straight-through jointing.

Having completed the conductor connection process the next stage is to complete the assembly of the joint enclosure before proceeding to the plumbing operation and subsequent filling of the enclosure with the appropriate compound.

One of the problems encountered in jointing is the fact that most joints are made in the open, often in inclement weather conditions, so that exclusion of moisture and dirt from the working area is of vital importance.

The jointing procedure for a paper-insulated lead-sheathed and armoured cable divides itself into five distinct sectors of operation:

1 Sheathing or serving and armour;
2 Lead sheath including plumbing (wiping) operation;
3 Insulation;
4 Conductors;
5 Joint enclosure.

Sheathing/serving and armour

Nowadays most power cables are sheathed with PVC and no special precautions need to be taken except where hydrocarbons may be encountered, such as in chemical factories and oil refineries, where it may become necessary to provide overall protection by means of an extruded lead sheath over the cable.

Served cable has changed little in construction over the years and generally consists of bitumenized paper tapes together with similarly impregnated hessian tapes having an overall coating of lime wash.

At jointing positions a PVC oversheath can be neatly trimmed at the point of removal and overtaped with a suitable PVC adhesive tape where necessary and the armour appropriately bonded.

In the case of hessian-taped serving this may be finished off at the joint entries by whipping with a suitable yarn or overtaping with a suitable hessian or fabric tape.

Lead sheath and joint enclosures

Cable sheaths are extruded from unalloyed lead, Alloy E (0.4% tin, 0.2% antimony) and Alloy B (0.85% antimony). Whilst unalloyed lead is generally in use for most armoured cables, Alloy E is used for unarmoured types and when vibration is expected in service. Severe cases of vibration require the use of Alloy B and it is recommended for cables running on bridges or suspended in catenaries.

The preparation of the wiping solder is of considerable importance and although the proportions of tin and lead vary between the individual preferences of jointers, it is generally accepted that two parts of lead (usually cable sheath) to one part of tin provides a good workable mix for wiping purposes at temperatures in the order of 250°C.

The plumbing metal is prepared by firstly melting some lead cable sheath in an iron pot to a temperature of about 300°C; about 2 ounces (57 g) of sulphur is added to the molten lead and mixed thoroughly, skimming off any dross which forms on the surface. More sulphur is added until the dross ceases to form. At this point about one ounce (28 g) of tallow is added and mixed thoroughly, and then one ounce (28 g) of resin is added and again mixed, after which the appropriate amount of tin should be added.

Tinman's solder, which is prepared in a similar way, but having between one and a half and twice the tin content of plumbing

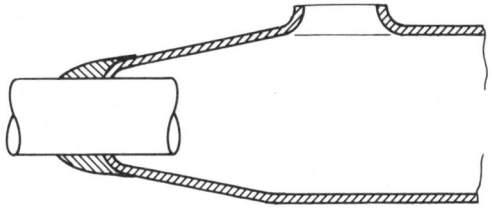

Figure 23 *Tapered-down lead sleeve*

metal, is used for sweating the conductor connections made by means of weak-back ferrules or lugs clamped onto the conductors and thoroughly basted with tinman's solder.

The process of soldering requires the use of a flux to prevent the oxidization of the cleaned metallic surfaces during the soldering operation. It is of importance to avoid the use of fluxes containing acid or salts; tallow is best for plumbing purposes whilst resin is appropriate for sweating operations on copper. A suitable resin paste may be made by mixing five parts of resin with two of palm oil.

Straight-through joints for paper-insulated cable generally incorporate a lead sleeve. If not of the compound sealed variety, this consists of a simple lead tube which needs to be dressed down onto the lead sheath of the cable by using a wooden tool known as a lead dresser (Figure 23).

After dressing, those parts of the sleeve where solder is to be applied are cleaned and tinned, with plumbers' black being applied to those areas of the sleeve adjacent to the tinned portions to prevent excess metal from adhering when plumbing takes place.

The plumbing metal is applied by the jointer from a ladle directly onto a wiping cloth and shaped to form a circular and bulbous protuberance at each end of the sleeve where it joins the cable sheath. A similar end result can be achieved by the use of a stick of solder and a gas torch or blowlamp subsequently shaping the solder after deposition by means of a wiping cloth. This particular technique is invariably used when a cable enters a dividing box in the vertical plane.

Insulation

The procedure for insulating conductors within the joint enclosure is, essentially, a straightforward process only requiring care and cleanliness on the part of the jointer and that the materials used are appropriate to the particular joint being made.

During the conductor connection operation the cable insulation should be protected by linen tapes which can be removed after the connections have been effected and the insulated cores made ready for appropriate additional insulation. This insulation can take the form of impregnated cotton or crepe paper tapes or shaped impregnated paper insulating tubes.

Stress control components may be required at the higher voltages but this subject is dealt with at length in another part of this book.

It is important to ensure that voids are not created during the insulating process as the presence of these can lead, at the higher voltages, to subsequent insulation breakdown.

Clearances to be maintained between phases and earth are given in Table 18.

The insulation and moisture-excluding process is completed by the introduction of compound into the joint enclosure.

A wide range of compounds exist with specially developed characteristics to suit specific operating conditions. Compounds generally used in paper cable terminations are bituminous based, although the use of cast resin as a filling medium has found some favour in place of the traditional paper cable filling compounds in recent years as has already been stated.

Having selected a suitable compound for the installation in question the compound should be heated until the recommended pouring temperature has been reached. The manufacturer's recommendation should be closely adhered to as it is important

Table 18 British standard clearances in terminations

Rated voltage (kV)	Insulating medium	Clearance between phases (mm)	Clearance between phase and earth (mm)
1.1	Compound or air	20	20
3.6	Compound	20	20
3.6	Air	90	65
12.0	Compound	45	32
24.0	Compound or oil	100	75
36.0	Compound or oil	125	100

that the compound flows freely but is not so hot as to damage the insulation of the cable. It is also important that the joint enclosure has been warmed thoroughly to ensure that on entering the enclosure the hot compound will not chill suddenly or unevenly; the danger of voids will then be obviated.

During the filling process pouring should take place in a continuous operation so that strata effects are avoided. The first pouring should fill the enclosure to approximately one third of its capacity; at this point stop pouring and allow about five minutes for the compound to fill the interstices of the joint, then fill to the two thirds level and wait a further five minutes before filling to the level of the filling hole, and top up as the compound cools but ensure that the pouring temperature is maintained throughout the topping-up process.

Conductors

Conductors may be connected either by soldering methods or by cold compression techniques.

The soldering or sweating process is quite straightforward and usually involves the use of a ferrule or copper sleeve.

Prior to the advent of the ferrule the technique of 'lapping' the conductors to be joined was very popular. The lapping technique involved the bending back of the outer strands of each cable core and butt-jointing the inner strands by soldering the butt. The outer strands were then brought forward over the butt, being cut alternately so that each outer strand made a butt connection with its opposite number. A copper wire binder, usually about 20 SWG, was then applied from 10 mm each side of the lapped conductors, the whole being sweated solid, smoothed with a file or glass paper and insulated with tape as required.

The ferrule consists of a copper tube which may be of the weak-back variety which can be clenched over the conductors prior to sweating.

Suitable-size ferrules are not always immediately to hand, particularly in an emergency situation, but it is often possible to manufacture a substitute on site if one is able to obtain suitable copper tube. In order to establish the length of the ferrule it is assumed that 100 amperes per square inch (645 cm^2) is a safe rating for soldered contacts.

Therefore the overall diameter of the ferrule D is calculated from the following formula:

$$D = \sqrt{\left(\frac{A + 0.785d^2}{0.785}\right)}$$

where A is the cross-sectional area of the conductor and d is its diameter.

The length of the ferrule L is obtained from the following:

$$L = \frac{2C}{100 \times 3.14d}$$

where C is the current carried by the conductor.

To permit an adequate flow of solder into the joint during the sweating process a slot or a row of holes should be provided in the ferrule (Figure 24).

Another form of soldered joint which achieved a high reputa-

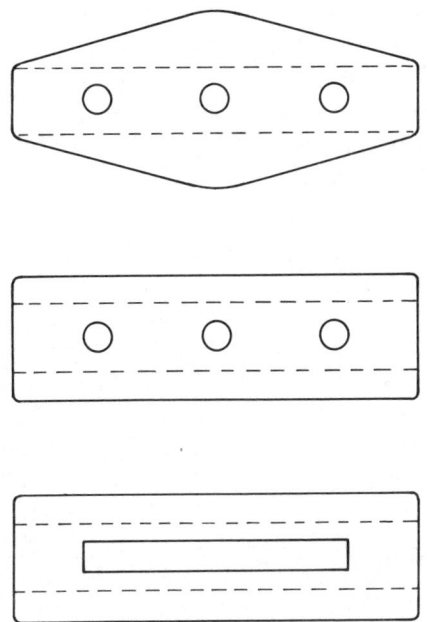

Figure 24 *Copper ferrules*

tion in the inter-war years was the Vernier braid joint which possessed high tensile strength as well as high conductivity. The joint consisted of strips of fine gauge copper braid which were laid lengthwise on the conductors and bound and soldered over the length. The flexible braids compressed closely into all the crevices between the wires, thus providing improved copper-to-copper contact. A further development in the braided joint was the vernier expansion joint, where the two conductors to be joined were allowed to slide within a brass tube, the electrical connection being made by means of copper braids sweated to the conductors beyond the end of each tube.

The advent of cold compression jointing using hydraulic tools, essential in the case of jointing aluminium conductors, removed the need for the use of soldered connections and, of course, overcame the limitation of 160°C conductor temperatures imposed under short-circuit conditions by soldered joints.

The compression connector consists of a lug or ferrule which can be compressed upon the conductor by a special die which either indents or circumferentially compresses the fitting upon the conductor by means of a hydraulically operated ram; this achieves good electrical conductivity together with high mechanical strength. Importantly, with aluminium it overcomes the problems associated with the troublesome aluminium oxide layer which quickly forms on the surface of the conductor.

The compression operation is quite straightforward and only requires the operator to ensure that the correct dies and fittings are selected for the conductor to be jointed.

PVC- and polymeric-insulated cables

The introduction of PVC mains cables resulting from the publication of BS 3346 in 1961, which regularized the use of plastic power cables up to and including 3.3 kV, provided users with a cable system which did not require the specialist services of plumber-jointers and removed the need to provide facilities for the heating of solder and compound.

The mining industry, where the need to joint paper-insulated cables in gaseous conditions had been a serious problem, was quick to see the advantages of PVC-insulated power cables and quickly evolved a series of specifications to meet the particular requirements of power transmission in coal mines.

At this time compression jointing was becoming available for conductor connections, and cold filling compound, a bituminous compound supplied in a liquid state to which a hardening agent was added, provided a completely heat-free jointing system.

The jointing operation associated with PVC-insulated mains cable is quite straightforward and does not require the degree of skill demanded by paper-insulated lead-sheathed cable. Essentially, neat removal of the PVC oversheath and insulation and sound connection of the conductors and armour bond is all that is required. This relatively simple operation does not require the degree of skill provided by the cable jointer and can be carried out by electrical workmen of the wireman category. The ease and speed with which this class of cable could be terminated led to the virtual demise of paper-insulated cable in low-voltage industrial applications during the 1960s.

In general, overground terminations of PVC-insulated mains cables do not require to be made within a compound-filled enclosure but where this is necessary, e.g. in flammable atmospheres or underground joint boxes, it is usual to employ a cold pour resin.

Stress control

The development of polymeric cables for use at voltages in excess of 3.3 kV brought about the need to control the electric field about the conductor. This was effected by the application of a semiconducting screen over the insulation which nullified the danger of electrical discharges occurring in voids around the periphery of the insulated core.

Similarly, stress points around the conductor itself need to be controlled by an extruded screen applied directly to the laid-up strands. Outer screens may consist of semiconducting tapes applied over a sprayed-on semiconducting paint, an extruded screen bonded to the insulation, or a strippable screen which, as the name implies, can be easily removed from the insulation when jointing takes place.

Before discussing the methods used to achieve stress control at power cable terminations, let us first consider the problems of high electrical stress on insulations at the cable connection position.

Figure 25a shows the concentration of electrical stress at the point where the earthed screen has been removed from the cable to

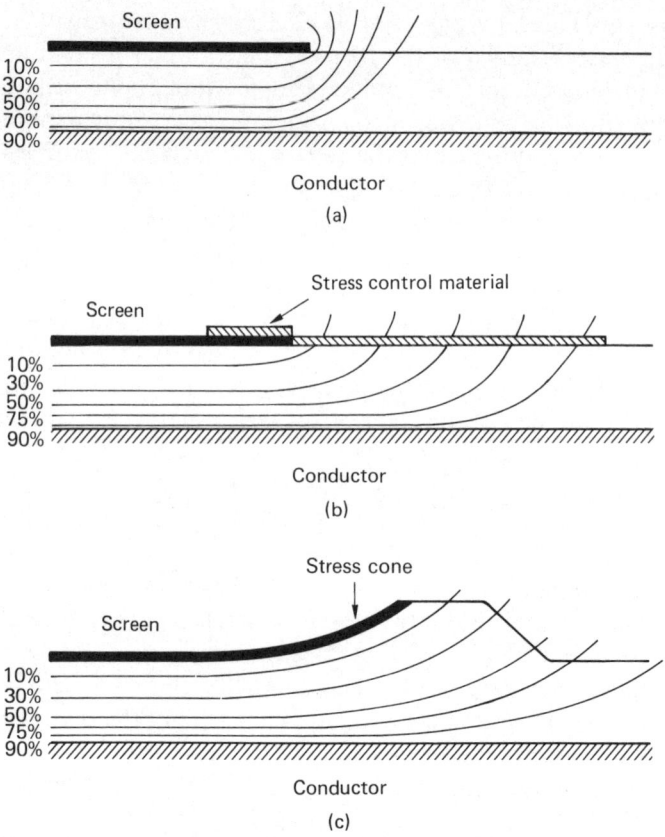

Figure 25 Stress control methods: (a) electric field at earthed screen termination; (b) electric field stress reduced by stress grading material; (c) electric field stress reduced by use of stress cone

provide the necessary clearance between phase and earth to prevent tracking occurring and consequent failure. It will be noted that a high stress point is established at the screen termination point and that this condition could lead to discharge and subsequent breakdown of the cable. The usual method of overcoming this problem on polymeric cables is to apply high-permittivity material to the screened core as shown in Figure 25b.

Semiconducting self-amalgamating tape is available which can be applied to the cable in a similar manner to conventional adhesive insulating tape. The function of this tape is to provide a resistive layer on the outside surface of the insulation, thus producing a resistive path which permits a smooth linear voltage gradient over its length.

Another technique makes use of pads of stress grading material which, after the screen has been removed, are applied to the cable over a band of quick-drying conductive paint. The pad is secured by self-amalgamating tape and works in a similar manner to the semiconducting tape stress control system.

The stress cone, which came into use originally with screened paper cables, works on the principle of controlling the capacitance at the point of termination of the screen; the cone continues past the point of screen termination, thus reducing the potential gradient to a safe level so that the possibility of ionization is avoided. Stress cones are used extensively in heat shrink and pre-moulded 'slip-on' power cable jointing systems. The principle of the stress cone is shown in Figure 25c.

Modern jointing techniques

Recent years have seen considerable innovation in the techniques and tools used in jointing. The advent of the polymeric cable has required a very different approach in the preparation of the cable prior to and during the physical process of making the electrical connection.

In all cable jointing operations cleanliness is of the utmost importance. Tools and the materials employed should be in good condition and suitable for their purpose. It is of vital importance when working on polymeric cables to strictly adhere to the jointing instructions provided.

It is particularly important that all contaminants and

Power Cable Installation Practice

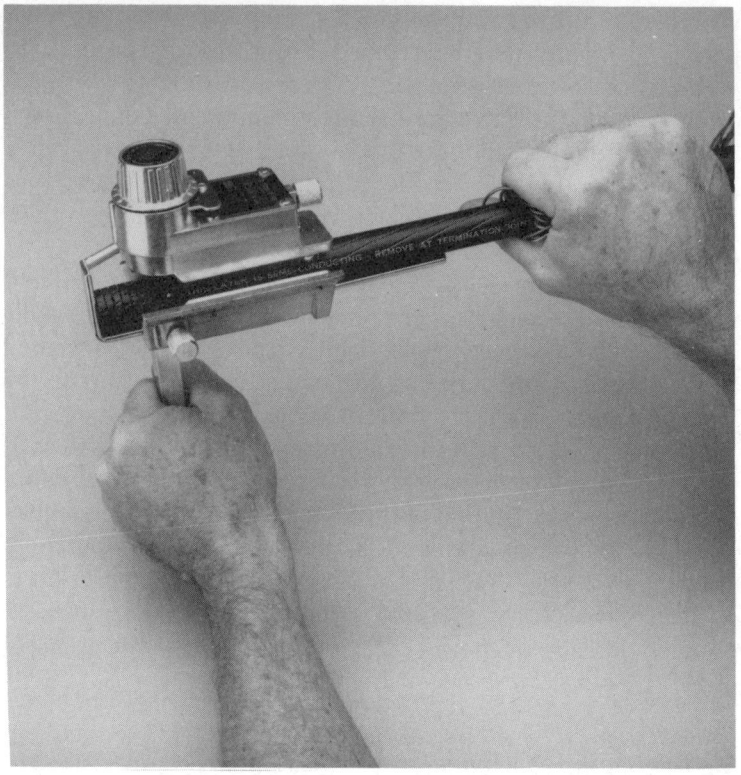

Figure 26 *SPEED insulation stripping and semiconducting screen scoring tools (Langley Engineering)*

semiconducting material *are completely removed from core screens and insulation* and that any discrepancies that may remain on the insulation are carefully rubbed down until clean and smooth with 25 mm wide silicon carbide (cloth back) abrasive strip using the following grades in sequence – 240, 320 and 400. In no circumstances should emery cloth ever be used.

The solvents generally used for cleaning screens and insulation

Power cable jointing

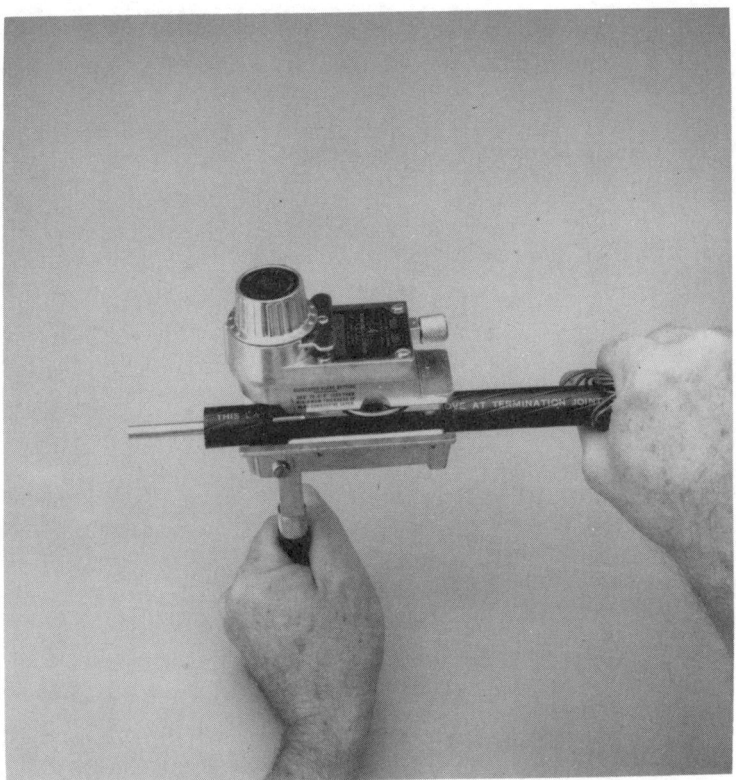

Figure 26 *continued*

are Genklene, Chlorothene or Inhibisol. When using these liquids care should be taken; use eye and skin protective equipment.

When cleaning screens and insulation use a clean white cloth dampened with the selected solvent, always wiping from the end at which the bare conductor is exposed towards the screen. It is important that the cloth should only be used once. The cleaning operation should be carried out as many times as necessary to ensure that the insulation and screens are *thoroughly* clean.

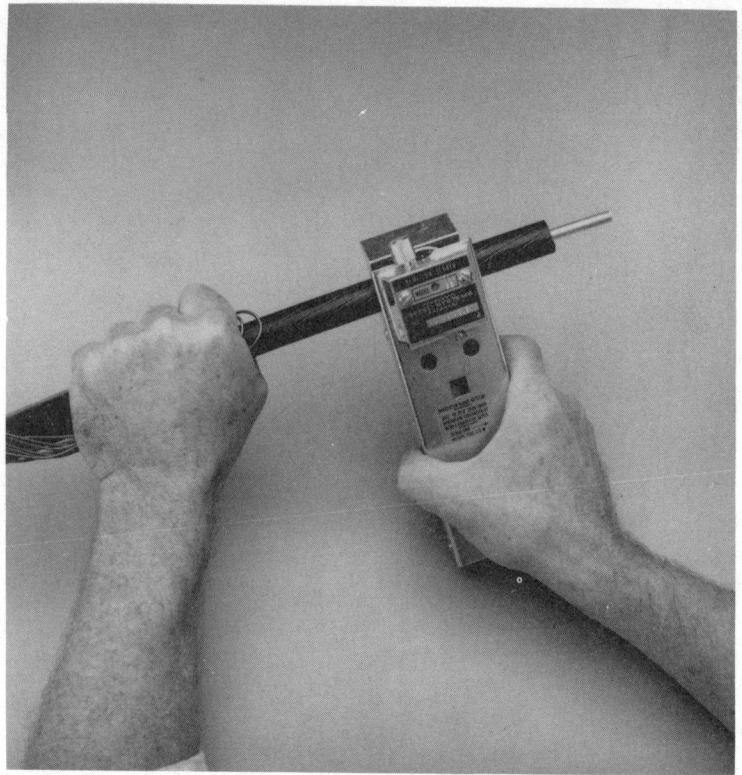

Figure 26 *continued*

Tools are available which enable the operator to remove both conductor insulation and semiconducting screens with a higher degree of accuracy than that obtained by using a knife. These tools can be set to close limits, thus producing consistent and precise terminations which cannot be achieved as easily by manual methods. The use of these specialized insulation strippers and screen removal instructions enables uniform results to be achieved throughout the cable jointing operation.

Power cable jointing

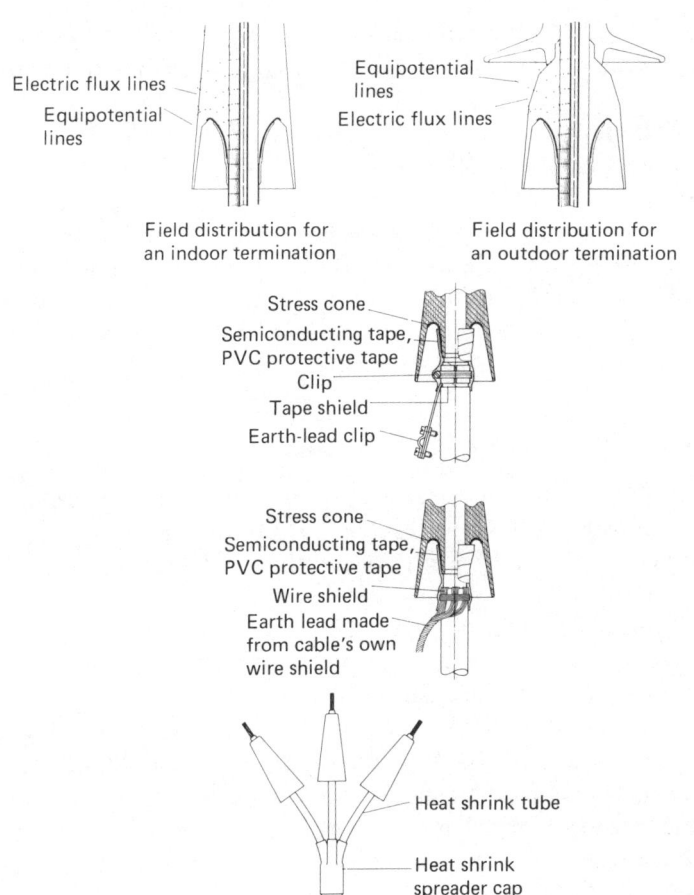

Figure 27 *Slip-on-type terminations (Cellpack (UK) Ltd)*

There are personal hazards in using many of the solvents currently in use and the instructions and cautionary advice provided should always be strictly adhered to.

The use of 'slip-on' components have come into increasing use in recent years. These are, essentially, elastomeric modules which can be simply pushed onto the prepared cable ends, thus dispensing with the need to enclose the joint or termination within a box or sleeve.

The modules have high elasticity which provides a high constant pressure which is maintained under all electrical loading conditions, thus avoiding the formation of voids between the insulation surfaces and the inside of the component. The modules are available for working voltages up to and including 33 kV.

Stress control is achieved by the inclusion of semiconducting material within the component which makes contact with the cable screen at the point of termination (Figure 27).

Handling power cables

Having selected a cable suitable for the service and the environment in which it is to be installed, it is important to ensure that damage is avoided by not attempting to bend it under low-temperature conditions.

Prior to installation the cable should be maintained at a temperature above 0°C for at least 24 hours before being removed from its drum for jointing or installation purposes. It is generally best to achieve this by keeping the drum in a heated building for this period; where this is not possible, the drum should be covered by a tarpaulin sheet, erected in the form of a tent, and carefully warmed by suitable oil or gas heaters. As this form of heat involves fire risks, the heating arrangements should be kept under supervision at all times.

Other precautions to be observed are to ensure that the cable oversheath has not been damaged and that the cable is not bent to a radius smaller than that recommended (Table 19).

Impregnated paper dielectric is hygroscopic, and it is therefore essential that lead-sheathed cable should always be sealed when not being worked upon. Unused cable after cutting must always be capped by plumbing on a lead cap immediately after the cutting operations.

Table 19 Minimum bending radius of power cables (BICC plc)

Type of cable	Overall diameter D	Minimum bending radius During installation	Minimum bending radius Adjacent to joints and terminations
BS 5467, BS6724 and BS6346			
Circular copper	10–25 mm	$4D$	$4D$
Conductor non-armoured	>25 mm	$6D$	$6D$
Circular copper conductor armoured	Any	$6D$	$6D$
Solid aluminium or shaped copper conductors armoured or non-armoured	Any	$8D$	$8D$
IEC 502			
Single core			
(a) Unarmoured	Any	$20D$	$15D$
(b) Armoured	Any	$15D$	$12D$
Three core			
(a) Unarmoured	Any	$15D$	$12D$
(b) Armoured	Any	$12D$	$10D$
BS 6480			
⩽6350–11 000 V			
(a) Single core	Any	$15D$	$15D$
(b) Multicore	Any	$12D$	$12D$
>6350–11 000 V and ⩽12 700–22 000 V			
(a) Single core	Any	$18D$	$18D$
(b) Multicore	Any	$15D$	$15D$

If it is suspected that moisture has entered a cable, through either a damaged lead sheath or cap, it is necessary to carry out a moisture test. This is carried out by heating about a litre of semi-fluid joint sleeve compound to 135°C in a suitable container. If compound is not available, melted paraffin wax may be used.

Samples of paper should be removed singly from layers nearest to and furthest from the conductor by means of tweezers to eliminate the possibility of moisture being conveyed to the paper from the hand. Immersion of moisture-contaminated paper in the liquid will cause it to bubble.

It will be necessary to cut back the cable and to remove and test paper samples until all traces of dampness disappear.

EIGHT
Cable testing and fault location

Reference to the appropriate British and IEC standards will give guidance with regard to the routine tests which are applied by the manufacturer to cables before they are despatched from the factory to the customer.

These tests involve the measurement of conductor resistance and insulation resistance, in accordance with the appropriate specification, together with any special tests which may be requested or deemed necessary to check manufacturing consistency or performance standards of the material used.

The application of high voltage will quickly find any defects in the insulation although it is of considerable importance not to apply an excessive pressure which could initiate long-term deterioration of the insulation and eventual failure of the cable in service.

Tests at the factory also include the measurement of the *dielectric power factor*, the dielectric loss angle (Figure 28).

The power factor of the dielectric of power cables needs to be kept at a low value, particularly so in the case of cables operating at 22 kV and above, and is given by the ratio:

$$\frac{\text{Loss of dielectric (watts)}}{\text{volts} \times \text{amps}}$$

The power factor of the dielectrics used in cable construction is very small, e.g. for paper cable in the order of 0.003; nevertheless it

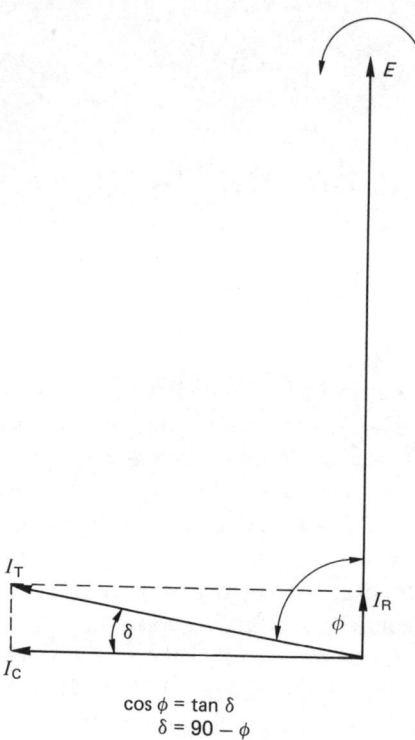

Figure 28 *Vector diagram representing loss angle*

represents a loss factor which becomes significant at the high-voltage range.

In a 'perfect' dielectric no loss occurs but a charging current I_C flows which is 90° in advance of the voltage E. However, in practice, a small current, I_R, flows in phase with voltage E. This current I_R causes losses in the dielectric which appear as heat.

The vectorial sum of these currents, I_C and I_R, is represented as I_T, which leads the voltage by an angle of less than 90°. The cosine of this angle, ϕ, is the power factor of the dielectric. With an efficient dielectric the angle ϕ is nearly equal to 90° so that $\cos \phi$ is approximately equal to $\tan(90 - \phi)$, i.e. equal to $\tan \delta$; so that the dielectric power factor of a cable is generally referred to as $\tan \delta$ where δ is known as the dielectric loss angle.

After site installation IEC standards require a DC test at a voltage level 70% of that applied during the routine factory tests. These pressure tests may be summarized as follows:

1 Paper-insulated cables up to 3.6–6 kV.
  Test applied for 5 min at $4.2U_0 + 3.36$ kV
2 Paper-insulated cables up to 6–10 kV and above.
  Test applied for 5 min at $4.2U_0$ kV
3 Polymeric cables.
  Test applied for 15 min at $4U_0$ kV

U_0 = Rated voltage between the conductor and earth.

It is of the utmost importance that these tests are carried out upon conclusion of the installation work to ensure that no damage has occurred during the laying and jointing operations.

Faults on power cables may be categorized into four main groups:

1 Insulation failure between conductors or between one core and earth.
2 Open-circuit on one or more cores with a high resistance path between the affected core or cores and other circuit elements at the fault location.
3 Low-resistance faults under open-circuit conditions.
4 Oil or gas leaks in impregnated pressure cables.

In order to localize a fault it is necessary to isolate the faulty circuit and ensure that the cores have been fully discharged to earth. The following tests should be made:

1 The insulation resistance is measured between each core and earth.
2 The insulation resistance is measured between cores.

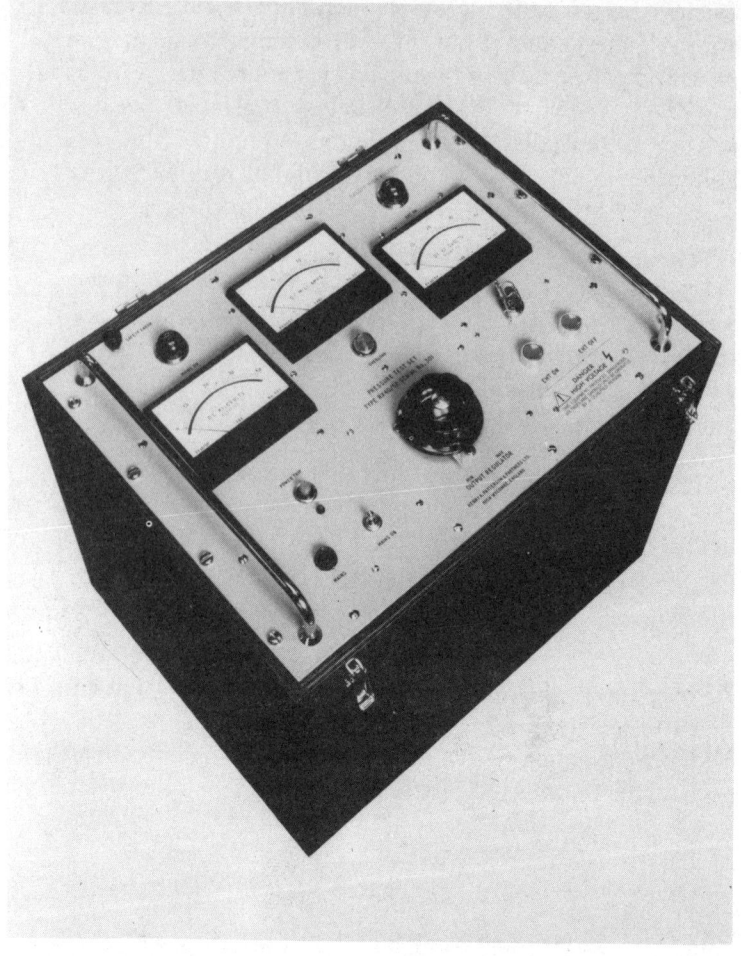

Figure 29 *Pressure test set (Henry A. Patterson & Partners Ltd)*

3 The loop resistance of the circuit is measured and compared with the original installation figures if available or, in the absence of these data, comparisons may be made with manufacturers' published figures.

From these tests it will be possible to establish the type of fault and classify its category which, in turn, will determine the type of localization test which will need to be applied.

A fault which falls into the first category can be, in most cases, tackled with the Murray loop test (Figure 30). Providing the resistance of the fault is relatively low and the 'sound' core has a reasonably high insulation resistance, one can obtain quite accurate results using the Murray loop test.

For guidance, if the resistance of the fault is in the order of 10 000 Ω to earth and the healthy core has an insulation resistance in the order of 1 MΩ, values of these levels should enable a satisfactory location of the fault to be made.

Figure 30 shows the circuit arrangement of a Wheatstone-type bridge used for location of (a) a core-to-earth fault and (b) a core-to-core fault.

In (a) the faulty core is bonded to a sound core at the remote end of the cable under test. In practice it is preferable to bond the cores directly to one another. If this is not possible, the interconnecting cable should be of the same cross-sectional area as the cores of the cable under test and its length added to the loop length.

The standard test set consists of decade resistors, a slide wire and a sensitive galvanometer; under balance conditions where a and b are decade resistors and portions of the slide wire, then:

$$\frac{a}{b} = \frac{x}{y}$$

or

$$a(a + b) = x(x + y)$$

Since resistance and distance are proportional and if $x + y$ is

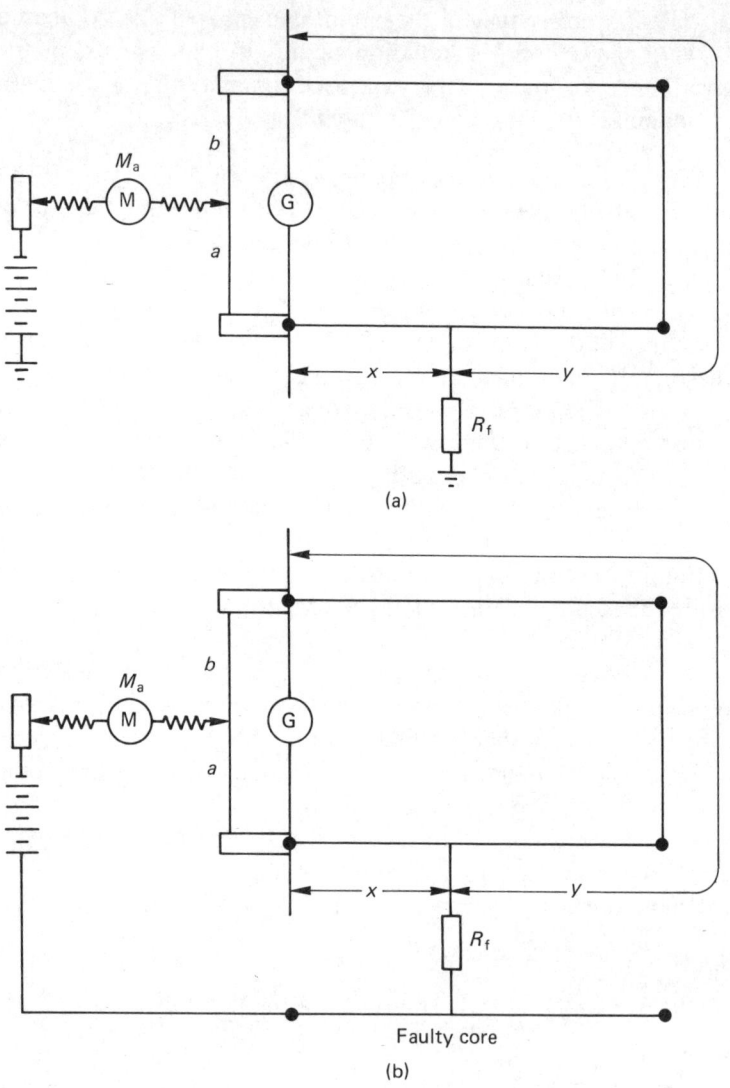

Figure 30 *The Murray loop test: (a) circuit for core-to-earth fault; (b) circuit for core-to-core fault*

Cable testing and fault location

taken as the loop length in metres then the fault distance X is given directly in metres:

$$x = \frac{a}{a+b} \times \text{loop length (m)}$$

For example, a cable 1000 m in length is looped at its distant end as shown in Figure 30a.

The bridge is balanced with the arm a at 50 Ω whilst b is set at 1000 Ω.

Then the distance from the testing position to the fault is found as follows:

$$x = \frac{50}{50 + 1000} \times 2000 \,\text{m}$$

$$= 95.24 \,\text{m}$$

Often it is not possible to pass sufficient current through the fault to arrive at a positive location. A current of at least 10 mA is required in the circuit to have a reasonable degree of certainty that a satisfactory fault position has been established.

If it is not possible to make a location using a standard bridge it is then necessary to 'condition' the fault by applying a high voltage from a DC test set to reduce the resistance of the fault to a level which will enable sufficient current to flow and thus achieve a balance of the bridge.

In ascertaining the loop test length it will be only twice the length if all the conductors forming the test circuit are of the same cross-sectional area. For example, if the faulty cable is 70 mm² and the return cable used for the loop is 95 mm² then the length of the loop will be:

$$1000 \,\text{m} + \left(\frac{70 \,\text{mm}^2}{95 \,\text{mm}^2} \times 1000 \,\text{m}\right) = 1737 \,\text{m}$$

If the cable making up the return section of the loop is of greater cross-section than the faulty one its equivalent length will be proportionately less than the route length and vice versa.

Power Cable Installation Practice

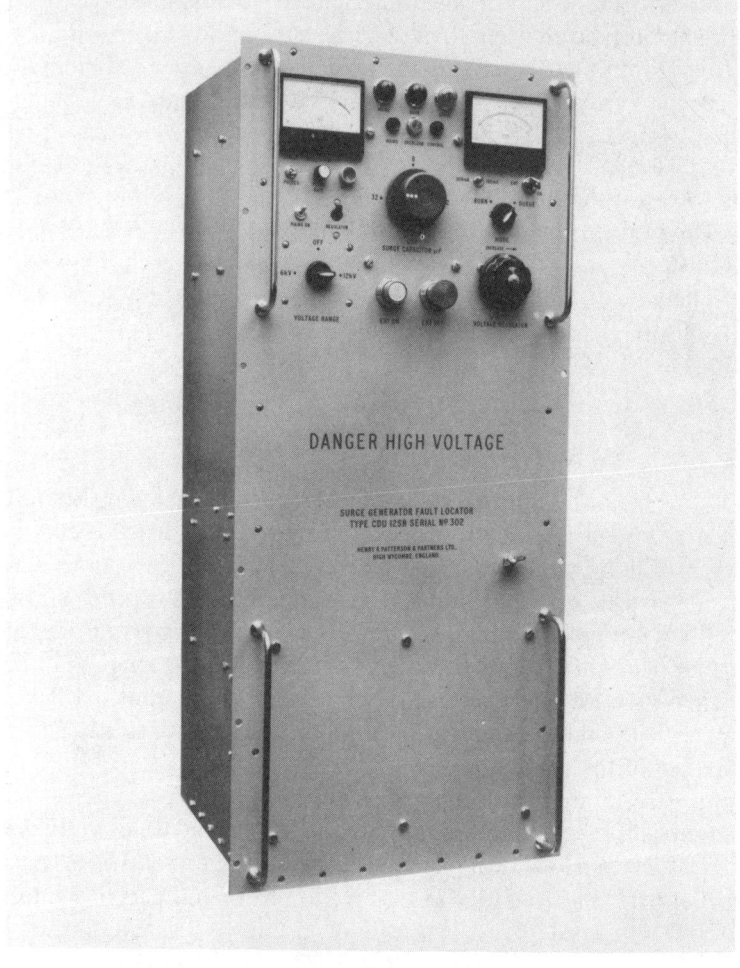

Figure 31 *Surge generator (Henry A. Patterson & Partners Ltd)*

Cable testing and fault location

The Murray loop test was the principal method used in power cable fault location for many years, but the need to restore circuits to operation in the shortest time possible led to the development of specialist equipment which enabled cable faults to be located quickly and accurately.

The technique of reducing the resistance of the fault, thus leading to more positive location, by the application of high voltage, the so-called 'Fault burning' technique, has already been mentioned. Suitable high-voltage DC test sets having the capability of providing an output voltage up to 80 kV at between 20 and 50 mA are readily available and will meet the needs of most small industrial users.

Nowadays, the electricity supply authorities, with extensive systems and obligations to maintain continuity of supply, have adopted more sophisticated techniques, such as the use of surge generators to either pre-locate or pinpoint the fault position by impulse current methods, or to detect the fault by seismic techniques, with an operator walking the cable route carrying a receiver which will respond to the acoustic disturbance caused by the flashover occurring at the fault position when a surge generator is connected to the faulty cable.

Such specialist equipment and personnel can often be hired by industrial users or contractors when faults are encountered which are beyond the scope of local resources.

A further recent technique used in cable fault location is the 'pulse-echo' method. The equipment used incorporates a pulse generator and a cathode ray oscilloscope. A signal sent down the faulty cable is reflected back and displayed on the screen of the cathode ray tube, together with other joints or branch connections. It is possible to identify the fault by analysis of the waveform and to measure the distance from the equipment to the fault which is displayed on the oscilloscope trace. However, it is not possible to precondition the fault with this type of equipment should a high-resistance or flashing-type failure be encountered.

There are numerous specialist fault-finding equipments on the

Power Cable Installation Practice

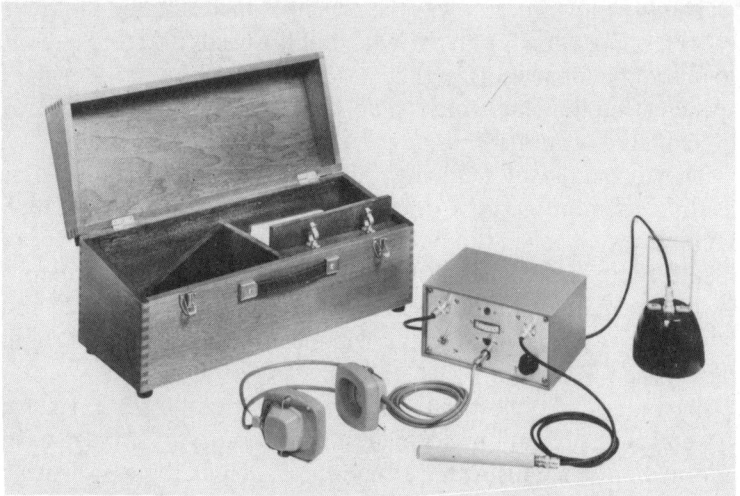

Figure 32 *Shock wave detector (Henry A. Patterson & Partners Ltd)*

market so that the user needs to determine his requirements relative to the specific features of his own installation or, in the case of a contractor, the class of work carried out. However, for systems up to and including 11 kV, a suitable selection of equipment would consist of a pressure and burn-out test set capable of producing 30 kV together with a 12 kV surge generator; also a pulse-echo set together with an acoustic detector and cable tracer.

The location of leaks in oil-filled or gas pressure cable systems, apart from those which occur in pipework and control panels, requires specialist techniques which can usually be provided by the cable manufacturer.

The usual procedure employed to locate an oil leak is to sectionalize the cable by freezing the oil, by means of liquid nitrogen, at a suitable section point. The oil pressure is then measured at joints on either side of the section point; this will establish on which side of the freeze the oil is escaping. This procedure is then repeated until a precise location has been established.

In the case of gas pressure cables the gas used is nitrogen, which is not easily identified by gas detection equipment, and it is therefore usual to introduce a tracer gas into the faulty cable. Sulphur hexafluoride (SF_6) is ideal for this purpose as it can be detected very easily.

The location of the leak is made by taking samples of the air at ground level immediately above the cable route and analysing these in a portable gas chromatograph.

If a power cable fails and a duplicate supply is not available, with consequent shutdown of plant and equipment, it is very often the best policy to bring in a specialist testing organization at the very outset in order to save time and consequential costs.

NINE
Earthing and bonding methods

Important features of any power cable installation are the earthing and bonding arrangements; the reader is therefore advised to consult the various Codes of Practice which have been published relating to the earthing and bonding requirements of specific industrial applications and also the relevant sections of the IEE Regulations for Electrical Installations in order to establish if any special or particular installation procedures have to be complied with.

The establishment of a suitable earth electrode system, usually at a main substation, requires specific consideration of the type of soil at the site of the proposed electrode. Firstly, the amount of moisture in the soil has an important bearing on the earth resistivity, which is usually expressed as ohm-centimetre or ohm-metre (1 ohm-metre = 100 ohm-centimetre). Because of the electrolytic nature of soil, conduction of current is determined by the degree of moisture content and the effect of soluble salts contained in the moisture together with the degree of compaction of soil and its grain size.

Seasonal variations can have a marked effect on moisture content and make it essential for the electrode to be installed at a depth which will maintain it in contact with permanently moist earth. Where this is not possible, which may happen in the case of earth plates, a number of earth rods distributed over an area is recommended.

Earthing and bonding methods

Table 20 Soil resistivity values (BICC plc)

Type of soil	Approximate value (Ohm-cm)
Ashes	350
Coke	20–800
Marshy ground	200–350
Loam and clay	400–15 000
Chalk	6000–40 000
Sand	9000–800 000
Peat	5000–50 000
Sandy gravel	5000–50 000
Rock	> 100 000

Reference to Table 20 will provide guidance to typical values of soil resistivity. Earthing arrangements are covered comprehensively and are detailed in Chapter 54 of the IEE Regulations for Electrical Installations (16th edn).

Some improvement of soil resistivity may often be obtained by treating the area to be used for earthing purposes with chemicals. Common salt is often used for this purpose.

In general, the length of earth rod used is of greater significance than the diameter as it will be apparent that longer rods can be driven to depths where the specific resistance is lower than nearer the surface. It is usual to connect the rod or plate electrodes together with a suitably sized conductor which is ultimately connected to a main earthing terminal and fitted with a disconnecting link in order that the earth system may be tested at regular intervals.

It should be noted that rods should, where possible, be spaced at not less than 2 m apart so that their respective resistance areas do not overlap the effective resistance area of an earth electrode extending some distance around the electrode (see Figure 33).

Under very heavy short-circuit conditions of long duration it is possible for dangerous voltages to exist on the surface of the

109

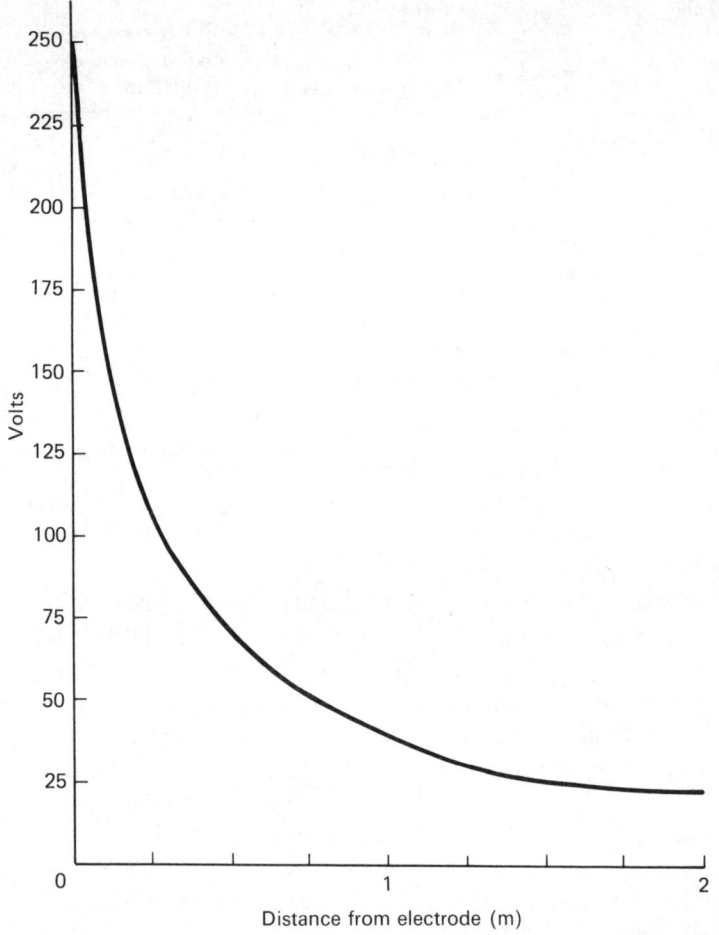

Figure 33 *Ground surface voltage measured in proximity to a copper rod electrode in clay/loam soil (BICC plc)*

Earthing and bonding methods

electrodes and it is therefore necessary that the appropriate requirements of the IEE Regulations for Electrical Installations are fully complied with.

It is also important to ensure that the conductors used for bonding, i.e. the connections between the various components of the system which are required to be maintained at earth potential, are adequate to provide sufficient thermal capacity and to contend with the heavy currents experienced during short-circuit conditions.

For guidance, it is recommended that 40 mm × 5 mm copper strip be used where short-circuit fault loads are expected to be between 75 and 150 MVA and 25 mm × 3 mm for installations below 75 MVA.

The metallic sheaths and armour of power cables can reach dangerously high potentials under conditions of insulation failure, fault current and electrostatic induction and it is, therefore, of extreme importance that lead sheaths and cable armour are solidly bonded together at the point where the connection to earth is made. Failure to observe good practice can lead to a voltage difference occurring at the connection point, leading to arcing and consequent damage to the lead sheath at this point.

Bonded cable systems

The use of large single-core paper-insulated lead-sheathed cables gives rise to problems associated with currents circulating in the lead sheaths. In order to minimize these circulating currents it is customary to install these cables so that they are in contact with one another; the trefoil formation is generally adopted (see Chapter 3). However, the trefoil configuration has the drawback of contributing to the mutual generation of heat between the three cables because of their close proximity to one another, although it must be pointed out that this problem does not become unduly significant until one reaches the larger conductor sizes and higher voltages.

It is necessary, as has already been pointed out, to bond and earth the sheaths of single-core cables in heavy-current circuits to avoid dangerously high voltages appearing on the sheaths, but, in turn, earthing the cable system at either end causes large circulating currents to flow in the lead sheaths which will limit the current rating of the cable system by reason of heating. If earthing and bonding is carried out at one point only on the system, sheath losses may be eliminated, but the cable sheaths at the end remote from the earthing point can acquire a voltage which may exceed the permissible value. This voltage is proportional to the conductor current; Figure 34 shows how cable spacing and sheath voltage are related.

In order to protect the sheath insulation from damage due to excessive sheath voltages, non-linear resistors known as 'sheath voltage limiters' are fitted at all joint or sealing end positions where the sheath is insulated from earth. These devices have a high impedance to current flow when the voltage is low but this collapses, thus providing a low-resistance path to earth, when the sheath voltage rises.

When a system is only earthed at one end it is known as an 'end-point bonded system' and all terminations, other than the one at which the earth connection is made, are insulated from ground and provided with sheath voltage limiters.

In order to cater for current flow under fault conditions an earth continuity conductor is provided between each end of the system. It is usual, in the UK, to limit sheath potential to 65 V and to shroud and insulate all terminations from contact with metalwork. Sheath voltage is proportional to the cable length and it is usual to limit end-point bonded systems to lengths not exceeding 500 m.

In the case of long cable routes it is often not possible to adopt an end-point bonded system and it is usual then to utilize the middle of the cable run to locate the earthing arrangements; this is then known as a 'mid-point bonding system'. Once again sheath voltage limiters are fitted at each termination position. It will also

Earthing and bonding methods

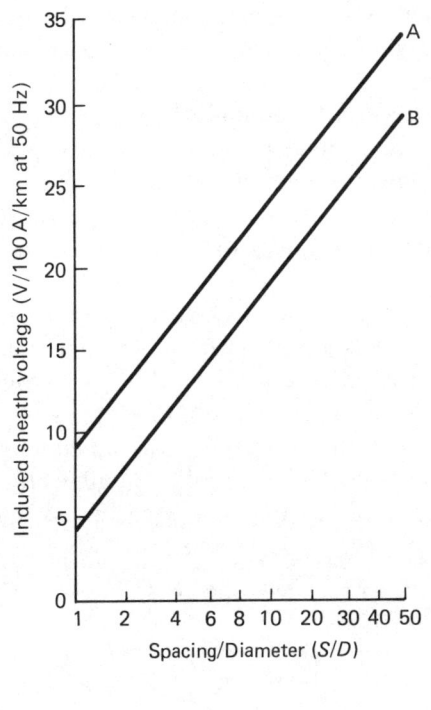

Figure 34 *Relationship of induced sheath voltage to conductor spacing (BICC plc)*

be seen that the sheath standing voltage may be allowed to rise to 65 V each side of the earth point when it is located in the middle of the cable route.

The technique of 'cross-bonding' is used for very long runs when it would not be technically acceptable to utilize the end or mid-point bonding systems. However, this method is generally only adopted for the higher voltage range of transmission cables where long circuit lengths and heavy current loadings pose particular problems.

By transposing the cable cores with respect to sheaths the three sheath voltages may be made to sum to zero, thus permitting earthing to be made at each cross-bonding point without the presence of circulating current.

In practice the route is generally divided into a series of major sections, each of which consists of three minor sections of equal length, usually one drum length of cable. Figure 35 shows a major

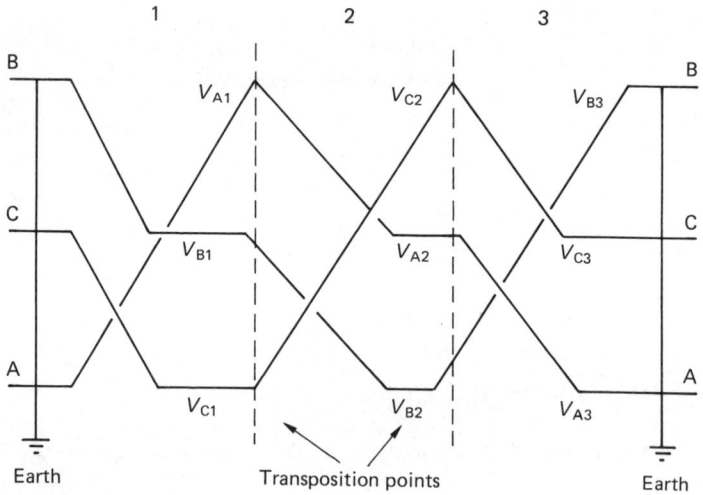

Figure 35 *Cross-bonding of three single-core cables, showing one major section comprising minor sections 1, 2 and 3. Phases are designated A, B and C. Voltages induced in each section are V_{A1}, V_{A2}, V_{A3} etc.*

Earthing and bonding methods

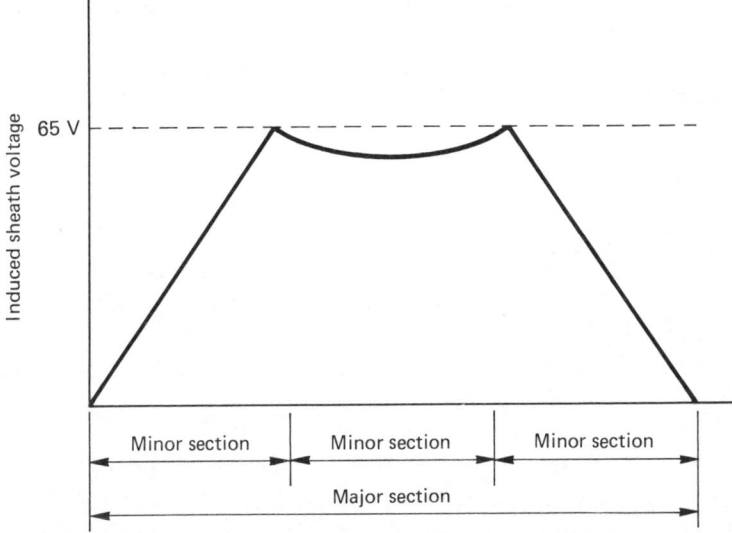

Figure 36 *Variation of voltage on a cross-bonded system sheath to earth (BICC plc)*

section with a complete transposition taking place over the three minor sections. At each end of the major section the sheaths are connected together whilst at each intermediate position they are transposed so that each major section consists of three sheath circuits, each of which is made up of three lengths of sheath, each of which is of a different phase conductor. If these sheaths are then cross-connected at the major section point the total voltage in each circuit will be the sum of the phase voltages at a displacement of 120 degrees and will therefore equate to zero. Reference to Figure 35 demonstrates the voltage disposition for one major section of a cross-bonded system. Once again the system is designed to restrict the sheath voltage to a maximum of 65 V over a major section with 0 V at the ends of each major section (Figure 36). Over long circuits the pattern is repeated and does not require

Power Cable Installation Practice

an earth continuity conductor as the sheaths over the whole cable route are continuously connected. Special accessories are available for the connection of sheaths at transposition and earthing locations.

Earth testing

In order to ensure the safety of an electrical installation it is necessary to measure the dissipation resistance of the earthing system at regular intervals as prescribed in Regulation 732-01-01 of the IEE Regulations for Electrical Installations.

The usual method applied for this measurement is that of the 'fall of potential method or alternating current injection'. The circuit used is shown in Figure 37.

An AC source supplies current to the electrode under test (T) and the auxiliary electrode (T_1) which should be separated by at least 50 m so that the resistance areas of the two electrodes do not overlap. The current injected is measured and also the fall in potential from the electrode under test (T) to the third electrode (T_2) situated at the mid-point between T and T_1.

The voltage drop between T and T_2 is then measured. The resistance of the earth electrode may then be found by noting the voltage between T and T_2 and dividing this figure by the current flowing between T and T_1.

In order to check the accuracy of the resistance value obtained, the second auxiliary electrode T_2 is moved first to position X and secondly to position Y with further readings being taken. If these results are generally in agreement the mean of the three readings may be taken as the resistance of the earth electrode T. If there is any discrepancy between the readings the tests should be repeated with the distance between T and T_1 increased.

The accuracy of the test depends upon an appreciable current being passed so that the rod used as an auxiliary electrode should be substantial enough to cater for the current employed.

A disadvantage in using this basic method of earth electrode

Earthing and bonding methods

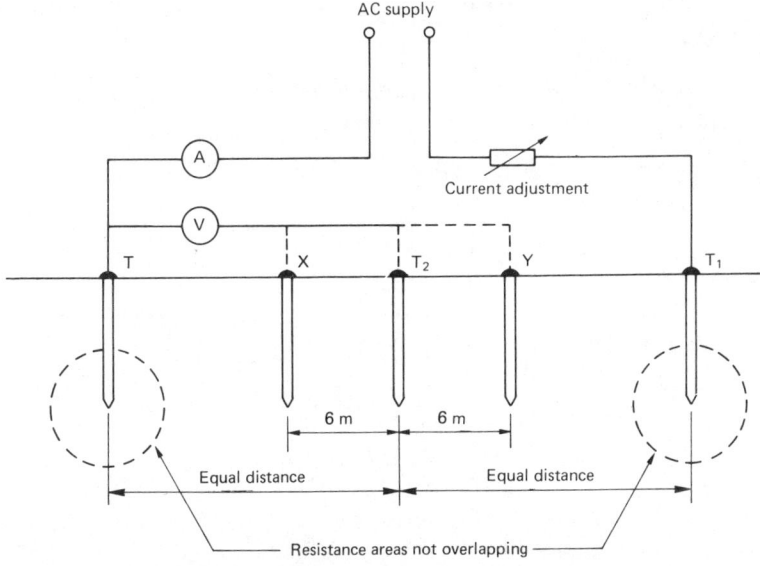

Figure 37 *Measurement of earth electrode resistance (IEE Regulations for Electrical Installations). T, earth electrode under test, disconnected; T_1, auxiliary earth electrode; T_2, second auxiliary earth electrode; X, alternative position of T_2 for check measurement; Y, further alternative position of T_2 for check measurement*

resistance measurement is that inaccuracies in potential measurement can occur due to couplings between the current- and potential-measuring circuits together with stray earth currents and EMFs due to electrolytic action in the soil. An earth testing ohmmeter eliminates some of these errors and has the advantage of being easily portable and independent of an external power supply. The instrument operates on reversed direct current, having commutators on the hand generator shaft which reverse the direction of current in both the control coil and the deflecting coil synchronously, measuring resistance only and being less susceptible to earth currents.

Other methods of earth resistance measurement sometimes used are:

1. The three-point or triangulation method
2. The ratio method

Details of the applications of these tests may be established by reference to appropriate literature on the subject.

TEN
Assisted-type cables

During the decade preceding World War I primary transmission voltages extended to 11 000 V and by 1914 cables operating at 22 000 V were beginning to be installed. However, as cable makers increased the voltage at which the paper-insulated cables of the period were designed to operate by the simple expedient of increasing the thickness of the paper insulation, it became evident that there were problems when operating voltages were extended beyond 22 kV.

In 1919 with the return to peace the demand for electricity increased dramatically and requirements arose for primary transmission cables operating at 33 000 V. By 1923 these cables, manufactured to pre-war standards, commenced to fail on a regular basis. Investigations by Martin Hochstaedter led to the introduction of a metallic screen around each power core which was maintained at earth potential, so reducing the levels of stress at the peripherals of the power conductors, and thus overcoming the insulation's propensity to failure brought about by the presence of high electrical stress in the coring up helix of the cable.

Whilst the introduction of core screens overcame the problems of electrical stress within the cable construction, the need to increase transmission voltages took the cable makers into further problem areas of design.

Cables incorporating Hochstaedter screens operating above 40 kV failed for seemingly unaccountable reasons. Examination of the failures revealed extensive carbon tracking on the papers adjacent to the point of failure and also the presence of a strange waxy deposit.

Investigations eventually led to the cause of the seemingly inexplicable failures which were, initially, due to the expansion of the impregnating compound, under the influence of cable loadings, forcing it through the paper layers towards the periphery of the cable and distending the lead sheath. During lightly loaded periods the cable cooled down but the compound, lacking restoring force and fluidic mobility, failed to return through the layers of paper so that partially vacuous spaces were created in the body of the dielectric in which ionization or flow discharges were initiated, the wax being a byproduct of this phenomenon.

Readers who wish to obtain more detail on this topic will find an article on the subject written by the author (Jones, 1989). For more extensive treatment of this subject the reader is directed to Robinson (1936).

In order to overcome the problems of voiding at high voltages the 'solid'-type cable could not be used at working pressures above 33 000 V and it was necessary to design a cable which incorporated a facility which would suppress any voids or gas pockets which might form in the dielectric.

There were two distinct approaches to the problem when it became apparent in the middle to late 1920s. Luigi Emmanueli, the chief engineer of the Italian Pirelli Company, had, during 1920, designed a cable which was filled with thin mineral oil that responded to volumetric change and restricted the formation of voids. The Emmanueli design was subsequently refined and the present-day oil-filled cables are almost identical to this original concept.

The second technique employed to overcome the problem of voids is to use gas pressure to suppress ionization, the so-called 'impregnated pressure' or IP cable or, alternatively, the gas pressure is applied to the dielectric via a flexible membrane of either lead or polyethylene; this external gas pressure cable is

Assisted-type cables

usually referred to as a 'gas-filled' or GF cable. In this type of cable the dielectric is maintained in a fully impregnated state by the application of external gas pressure to the membrane, thus preventing the formation of voids. Because this type of cable requires both a membrane and a pressure-resistant sheath it is generally uncompetitive with other pressure-assisted cables.

Oil-filled cable systems

The oil-filled system based on the original Emmanueli design has been highly successful and is now the most widely used solution to the problems encountered in transmitting power at voltages in excess of 33 000 V underground.

The oil-filled cable consists, essentially, of oil-impregnated-paper-insulated conductors enclosed in a hermetically sealed metal sheath capable of withstanding internal pressures of up to $524 \, kN/m^2$.

The fundamental design principle enshrined in the oil-filled cable is that the metal sheath should be completely filled with oil. This is achieved by providing oil channels within the cable which allow low-viscosity oil, which expands and contracts under the effect of temperature variations produced by the service load cycle of the cable, to flow to and from hermetically sealed oil reservoirs or 'presure tanks' directly connected to the cable at points along its route. Figure 38 shows a diagrammatic layout of such a system.

In designing an oil-filled cable system, hydraulic factors must obviously be taken into account, and the following parameters are usually accepted as appropriate: 5.25 bar for sustained pressure and 8 bar for transient pressure. The minimum design value is usually accepted as 0.2 bar.

A specially designed range of accessories is required for oil-filled cable systems. These can be detailed as follows:

Straight joints   For connecting lengths of cable.

Power Cable Installation Practice

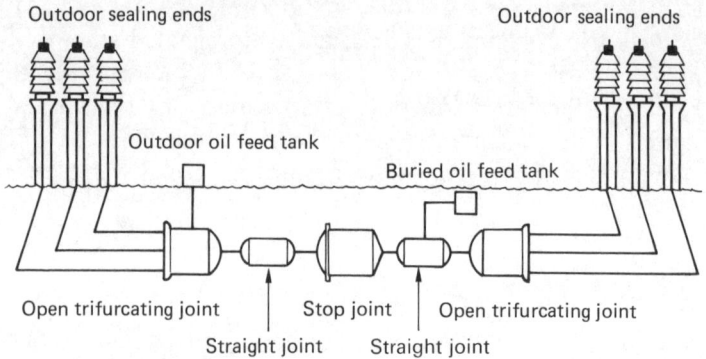

Figure 38 *Typical three-core oil-filled cable system*

Trifurcating open joints For connecting a three-core cable to three single-core cables.

Straight stop joints For dividing the cable system into separate oil sections, because of either the length of the circuit or the profile of the route.

Trifurcating stop joints For connecting a three-core cable to three single-core cables which may be of the solid type in the case of a 33 kV system.

Outdoor sealing ends For terminating cables outdoors.

Oil-immersed sealing ends For terminating cables directly into the terminal chambers of switchgear and transformers, thus facilitating the design of more compact apparatus.

Oil reservoirs The oil reservoir or 'pressure tank' consists of a closed cylindrical tank encasing a number of flexible-walled metal cells. These cells are filled with a gas which may be at atmospheric pressure or higher depending upon the particular service for which the tank is required. The space within the tank not occupied by the cells is filled with degasified oil which takes up between 30 per cent and 50 per cent of the nominal capacity. The tanks are connected by pipes to the oil ducts in the cable.

In service any increase in cable temperatures will result in oil being expelled from the cables into the pressure tank, where it compresses the cells, thus increasing the pressure within the tank. This increased pressure is then available to force the oil back into the cable when the cable temperature falls.

The tanks are located on the cable route at points determined during the hydraulic survey at which time the size of individual units and the pressurization levels are established.

Generally, up to about 150 kV, three-core oil-filled cables are standard but from 200 kV upwards single-core cables are generally most suitable.

Gas pressure cables

The impregnated pressure (IP) as distinct from the gas-filled (GF) cable has been preferred for service in the UK largely due to the pioneering efforts of the former Callender's Cable and Construction Co. Ltd, who installed the first 132 kV circuit at Burford in Gloucestershire in 1940. However, for the last few decades, the oil-filled cable has proved to be a more attractive proposition although there are installation conditions which would often favour an IP cable above an oil-filled one.

The cable is constructed in a conventional manner, being sheathed, armoured and protectively served according to the requirements of the installation. In the case of lead-sheathed gas-filled cables, impregnated and drained jute strings are laid into the interstices to allow gas to penetrate along the cable length in contact with the pre-impregnated paper insulation. Aluminium-sheathed gas-filled cable may be supplied without ducts or fillers in the interstices.

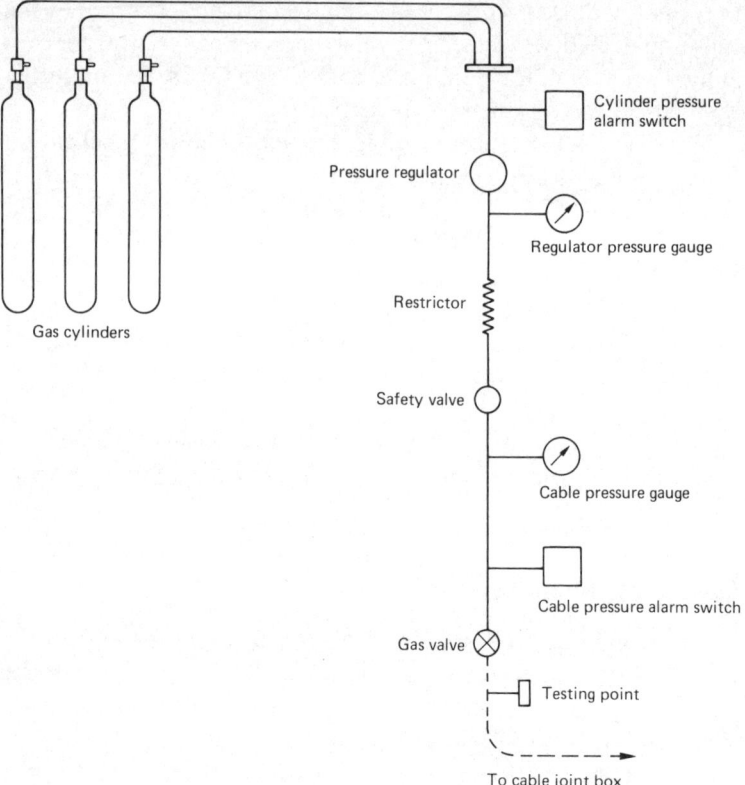

Figure 39 *Control panel gas connections for three-core impregnated pressure cable*

After installation the cable is charged with nitrogen at a nominal pressure of 14 bar which is controlled from a cubicle located at the end of the feeder. Although the cable and its accessories are designed for gas-tightness under normal operating conditions, means are provided for maintaining adequate pressure in the system should a gas leak occur due to mechanical damage or some other cause.

The cubicle houses a number of nitrogen cylinders, normally connected to the cable through a pressure-reducing valve (Figure 39) together with a pressure alarm switch which gives warning should the gas pressure fall below 12.4 bar.

Operation of the alarm system does not necessarily mean that the cable must be taken out of service immediately but rather that an inspection is required. Even with a severe leak, such as a pick hole in the sheath, the cable can be kept in service by replacing the gas cylinders as they become exhausted. Techniques suitable for leak detection are described in Chapter 8.

In the years that gas-filled cables have been in operation they have given excellent service. The terminal equipment is simple, and bottled nitrogen is readily available in most parts of the world, so that it is possible in more remote areas to keep the cable in service for some time until the services of a specialist supertension cable jointer become available.

Although the cost of cable construction for a gas-filled cable, because of heavier metal sheaths and lower electrical design stress, generally makes IP more expensive than the equivalent oil-filled cable there are, nevertheless, considerable savings in reducing jointing costs, absence of stop joints etc. which can often give an IP system the economic edge, if fully costed, over its rival.

References

Jones, E. W. P. (1989) 'The great dielectric phenomenon', *Power Engineering Journal*, 3, No. 1.

Robinson, D. M. (1936) *Dielectric Phenomenon in High Voltage Cables*, Chapman and Hall.

Index

Accessories, 66
Aluminium, 20–22, 58
American Wire Gauge, 19
Armour, 50, 79

Bend radius, 95
Bentonite, 24, 40
Bonding, 111
Breakdown strength, paper, 8
Butyl rubber, 12

Calcium carbonate, 12
Callender, W.O., 3
Circular mil, 19
Cleat installation, 29
Cold compression, 20, 71, 83
Compound, 70
CONSAC, 22
Continuous current rating, 33
Cotton, 7
Cross bonding, 114
Crosslinked polyethylene, 14
Current:
   continuous, 33
   fault, 33

Dielectric power factor, 97
Direct laying, 26
Dividing box, 67
Double steel tape armour, 21
Duct, 24
   installation, 27

Earth electrode, 109
Earth fault, 50
   Paper, 55
   PVC, 56
   XLPE, 57
Earth testing, 116
Elastomer, 2, 76
Emmanueli, Luigi, 120
End point bonding, 112
EPR, 14, 17, 49, 77
ERA Technology Ltd, 33
Ethylene propylene rubber, 14

Faraday, M., 2
Fault burning, 105
Fault current, 47
   calculation, 60
   earth fault, 59
   short circuit, 59

Index

Ferranti, S., 6
Ferranti, S.Z., 66
Ferrule, 83
Flame propagation, 22

Gas pressure cables, 123
Gas-filled cable, 120
Ground surface voltage, 110
Gum rosin, 9
Gutta-percha, 2

Hochstaedter screen, 119
House service cable, 1
Hydrogen chloride, 12

IEE Regulations, 24, 30, 40, 42, 44, 48, 59
Imperial System, 19
Impregnated paper, 1, 6
Insecticides, 23
Insulation failure, 99

Loss angle, 98
Losses, 14
   armouring and sheath, 34
   eddy current, 34
Low smoke and fume, 13
Low-resistance faults, 99

Manilla, 7
Metric system, 19
Mid-point bonding, 112
Milliken, 20, 54
MIND cables, 9, 10

Mineral oil, 9
Moisture test, 96
Monosil process, 14
Murray loop test, 101

Oil-filled cable, 121
Open circuit, 99
Ozone, 11

Paper-insulated lead-sheathed, 1, 21, 78
Permittivity, 12
Plastic tape, 23
Polyethylene, 2, 13
Polyvinyl chloride, 2, 12
Power factor, 12
Pre-arcing, 47
   let through energy, 47
Protection, 21
Pulse-echo method, 105
PVC, 12, 28, 34, 50, 77

Rack installation, 28
Reactance, 61
Resin, 73
Rodents, 22
Rubber, 1
   natural, 10
   synthetic, 11

Serving, 79
Sheath voltage limiter, 112
Sheathing, 79
Short circuit, 46
   conductor temperature, 14

Index

Short circuit rating:
   paper, 51
   PVC, 52
   XLPE, 53
Short-circuit current,
   calculation, 59
Slip-on termination, 76
Soil resistivity, 108
Soil thermal resistivity, 38
Soldered joint, 58
Soldering, 20, 83
Steel wire armour, 21
Straight joint, 66
Strees:
   cone, 88
   control, 87
   spacing of, 32
Supports, 104
Surge generator, 67

T-joint, 67
Temperature, 94
   air, 4, 36
   ambient, 36, 42
   conductor, 37, 39
   ground, 36, 39
   rise, 49
   short circuit, 50
Terylene, 4, 6
Tile, warning, 23
Toxic gases, 22
Tray installation, 30
Trefoil:
   cleat, 29, 32
   duct group, 41
Trench depth, 23, 41

Varnished cambric, 4
Voltage drop, 42
Vulcanized bitumen, 1, 3

Warning boards, 23
Water treeing, 18
Wiping, 70, 80
Wood pulp, 7

XLPE, 14, 17, 28, 34, 49, 50, 77